Recent Progress in Separation of Macromolecules and Particulates

ACS SYMPOSIUM SERIES **1281**

Recent Progress in Separation of Macromolecules and Particulates

Yongmei Wang, Editor
Department of Chemistry, The University of Memphis
Memphis, Tennessee

Wei Gao, Editor
Analytical Sciences, The Dow Chemical Company
Collegeville, Pennsylvania

Sara Orski, Editor
Materials Science & Engineering Division
National Institute of Standards and Technology
Gaithersburg, Maryland

X. Michael Liu, Editor
Materials & Product Chemistry, Pfizer Inc.
Richmond, Virginia

Sponsored by the
ACS Division of Polymer Chemistry, Inc.

American Chemical Society, Washington, DC

Distributed in print by Oxford University Press

Library of Congress Cataloging-in-Publication Data

Names: Wang, Yongmei, editor. | Gao, Wei, 1969- editor. | Orski, Sara, 1984- editor. | Liu, X. Michael, editor.
Title: Recent progress in separation of macromolecules and particulates / Yongmei Wang (University of Memphis), Wei Gao (Dow Chemical Company), Sara Orski (National Institute of Standards and Technology), X. Michael Liu (Pfizer Inc.), editors.
Description: Washington, DC : American Chemical Society, [2018] | Series: ACS symposium series ; 1281 | Includes bibliographical references and index.
Identifiers: LCCN 2018033950 (print) | LCCN 2018049635 (ebook) | ISBN 9780841233089 (ebook) | ISBN 9780841233096
Subjects: LCSH: Separation (Technology) | Macromolecules. | Chromatographic analysis.
Classification: LCC TP156.S45 (ebook) | LCC TP156.S45 R45 2018 (print) | DDC 660/.2842--dc23
LC record available at https://lccn.loc.gov/2018033950

Distributed in print by Oxford University Press

Foreword

The ACS Symposium Series was first published in 1974 to provide a mechanism for publishing symposia quickly in book form. The purpose of the series is to publish timely, comprehensive books developed from the ACS sponsored symposia based on current scientific research. Occasionally, books are developed from symposia sponsored by other organizations when the topic is of keen interest to the chemistry audience.

Before agreeing to publish a book, the proposed table of contents is reviewed for appropriate and comprehensive coverage and for interest to the audience. Some papers may be excluded to better focus the book; others may be added to provide comprehensiveness. When appropriate, overview or introductory chapters are added. Drafts of chapters are peer-reviewed prior to final acceptance or rejection, and manuscripts are prepared in camera-ready format.

As a rule, only original research papers and original review papers are included in the volumes. Verbatim reproductions of previous published papers are not accepted.

ACS Books Department

Contents

Indexes

Preface

Separation and characterization of materials is an essential step in the advancement of science and technology. The need for separation and characterization could arise from wishes to obtain purified samples with less contaminants, requirements for quality control of materials for downstream applications, or needs to identify structures and conformations of active species present in the applications. The types of materials encountered in applications are diverse, from synthetic polymers and biological derived biopolymers to micro or submicro particles and various conjugated or self-assembled nanomaterials. This makes the task of separation and characterization challenging. There is no one-size-fits-all approach. Scientists, whether working in academia or in industries, are constantly facing new challenges. Communication between scientists is essential for keeping each other updated on developments in the field. This was the motivation for organizing the symposium on "Recent Progress in Separation of Macromolecules and Particulates" at the 2017 Spring National American Chemical Society (ACS) Meeting in San Francisco. Symposiums based on such a theme have been organized several times in ACS meetings as well as in other settings, but no symposium books have been published. Hence, we decided to put together this symposium book, based on the talks presented at the symposium, in the hope that knowledge shared at the meeting can be shared more broadly within the community and beyond.

This book begins with a chapter by Taihyun Chang on "Chromatography Separation of Polymers", which gives a nice introduction to polymer liquid chromatography methods for characterizing polymers with complex architectures and chemical compositions. Chang discusses how Interaction Chromatography (IC), especially Thermal-Gradient Interaction Chromatography (TGIC), provides far better separation resolutions for polymer samples. Coupling of IC with Size Exclusion Chromatography (SEC) or another separation chromatography leads to two-dimensional (2D) chromatography that provides even better resolution. Chang gives several examples of high resolution separations of complex polymer systems achieved in his lab. These examples should provide valuable insight into the capability of utilizing various polymer liquid chromatography modes for separation and characterization of polymer samples.

This is followed with a chapter by Yongmei Wang on "Theory and Principles of Interaction Chromatography" that explains the physical origin behind various polymer separation chromatography modes. Wang and her group used computer simulations to discuss and illustrate the principle behind Liquid Chromatography at the Critical Condition (LCCC). This chapter should provide readers with a solid theoretical understanding of polymer adsorption and how polymer adsorption plays out in the LCCC mode. The next chapter (Chapter 3), contributed by Han and Ryu, illustrates how one can remove homopolymer species present in as-synthesized block copolymers, prepared by anionic polymerization, via polymer adsorption on silica nanoporous materials. The success of such an approach hinges on phenomena related to polymer adsorption encountered in Interaction Chromatography.

The next two chapters are contributed by scientists working in industry, which each has its own challenges and needs. The first (Chapter 4), contributed by Crawshaw et al, discusses how to characterize gelatin samples derived from biological sources. These authors faced the challenge of learning as much as possible about the gelatin samples in terms of the molecular weight distribution and conformations of the chains. These authors used gel permeation chromatography (GPC) coupled with advanced multi-detectors and showed how one can extract information that reflects molecular weight distributions and chain conformations. The next chapter (Chapter 5), contributed by Zhang et al, discusses the challenge of identifying oligomer species present in the aqueous phase during emulsion polymerization. These authors used LC-MS to identify the oligomer species present and also showed how one might obtain logD values from LC-MS to learn about the interfacial properties of these oligomer species.

Field Flow Fractionation (FFF), developed by J. Calvin Giddings and his colleagues, has become another important approach for separation and characterization of macromolecules and particulates. Unlike column chromatography which utilizes packing materials to achieve separation, FFF utilizes a field applied to a fluid or solution to achieve separation. FFF is known to be more suitable than GPC for characterizing polymers with high molecular weights. But before we present chapters that utilize FFF technique, we first have a chapter contributed by Meunier et al that pushes the limit of GPC in characterizing ultralarge polymers. These authors discuss several parameters that can be adjusted in GPC experiments, such as flow rate, column diameter, column temperature and packing particle size, that can push the limit of GPC in characterizing polymer samples with ultrahigh molecular weights. This chapter (Chapter 6) should provide readers with a good understanding of the factors that could limit the separation achieved in GPC.

This is by followed by several chapters that use FFF, especially Asymmetric Flow Field Flow Fractionation (AF4), to characterize particles and biomacromolecules. The chapter (Chapter 7) contributed by Gao et al illustrates the sensitivity of AF4 coupled with different detectors to obtain particle size distributions for silica colloidal particle samples, especially the sensitivity to identify lot-to-lot variation in particle size distributions. The chapter (Chapter 8) contributed by Li et al compares AF4 with SEC in characterizing biopolymers such as bovine serum albumin (BSA) and immunoglobulin (IgG). While AF4 and SEC both provide consistent size distributions, AF4 is shown to be uniquely suited for biopolymers that may form aggregates. Last, the chapter (Chapter 9) contributed by Lederer et al shows several examples where AF4 is utilized to characterize synthetic macromolecular systems encountered in drug delivery systems.

We thank all the authors for contributing to the book and all the reviewers for reviewing chapters. We thank Dr. Jesse Ziebarth for proof-reading all the chapters. Without these efforts, this book would not be possible.

Yongmei Wang
Professor
Department of Chemistry
The University of Memphis
Memphis, Tennessee 38152

Wei Gao
The Dow Chemical Company; Analytical Sciences
400 Arcola Road, Collegeville, Pennsylvania 19426

Sara Orski
National Institute of Standards and Technology
Materials Science and Engineering Division
Polymers and Complex Fluids Group
100 Bureau Drive, Gaithersburg, Maryland 20899

X. Michael Liu
Pfizer Consumer Healthcare
P.O. Box 26609, Richmond, Virginia 23261

Chapter 1

Chromatographic Separation of Polymers

Taihyun Chang*

**Department of Chemistry and Division of Advanced Materials Science,
Pohang University of Science and Technology,
Pohang, 37673, Korea
*E-mail: tc@postech.ac.kr.**

Polymers are usually mixtures of molecules that are inhomogeneous in various molecular characteristics. For precise characterization of polymers, distributions in all of these molecular characteristics need to be addressed, which is generally an impossible task. However, it is often sufficient in practice to analyze a limited number of molecular characteristics. Liquid chromatography is the most widely used tool for this purpose and has made remarkable progress during the last half century in both instrumentation and understanding of the separation mechanism. In this chapter, various chromatographic techniques developed for polymer separation are briefly reviewed.

1. Introduction

Except for some biopolymers, most synthetic and natural polymers are not homogeneous in various molecular characteristics such as molecular weight (MW), chemical composition, chain architecture, etc. Distributions in the molecular characteristics are reflected in the final properties of polymeric materials and the precise characterization of these distributions is important in design and performance control of the polymeric materials. To analyze the multivariate distributions of a polymer sample, it is desirable to separate the polymer sample first according to the molecular characteristics of interest. For this purpose, liquid chromatography (LC) is the most commonly used technique.

Among numerous variations of the LC methods, size exclusion chromatography (SEC) has been used most widely for molecular weight distribution (MWD) analysis of polymers (*1–4*). SEC separates polymer analytes using the distribution equilibrium of polymer chains between a common solvent in the pore and the interstitial volumes of the LC column packed with porous spherical particles. In an ideal SEC separation, the column packing materials should not adsorb the polymers and the distribution equilibrium of the polymers is solely due to the entropic exclusion of polymer chains from the pores (*5*). Therefore, the SEC eluent is commonly a strong (promoting desorption) solvent for the polymer analytes. (The solvent strength is a different concept from the thermodynamic solvent quality but a strong solvent is generally a good solvent.) SEC separation has advantages over other classical fractionation methods in speed and the amount of the sample required. SEC separation can be combined with a variety of detectors such as light scattering and viscosity detectors that allow measuring absolute MW and intrinsic viscosity online in conjunction with concentration detectors such as UV absorption or differential refractive index detectors.

Nonetheless, it is important to keep its limitation in mind. First, SEC separates the polymers according to their hydrodynamic size. While the separation being sensitive to one molecular characteristic is of course a distinct merit, as it is the principle of the universal calibration in SEC (*6*), SEC separation cannot distinguish between other molecular characteristics such as chain architecture or chemical composition beyond the chain size in the eluting solvent. Secondly, the resolution of SEC is limited due to the band broadening of the separation method. Although band broadening is a general phenomenon in chromatography, it needs to be considered carefully in the MWD characterization of polymers since the elution peak of a polymer sample is a convolution product of MWD and the instrumental band broadening (*7*).

To complement the limitation of SEC, another LC method utilizing the enthalpic interaction of analytes with the stationary phase, called liquid adsorption chromatography or interaction chromatography, has been increasingly used in recent years (*8–13*). In principle, it is nothing but a typical HPLC method widely used for the separation of small molecules and the term of interaction chromatography (IC) will be used in contrast to SEC in this chapter. Since the IC separation is based on the interaction of polymer analytes with the stationary phase (surface of the packing materials in the column), the distribution equilibrium of the analytes between the mobile and the stationary phases is sensitive to not only MW but also the chemical nature of the polymers. The eluents used for IC separations are generally weak (promoting adsorption) to induce appropriate interaction strength. In terms of instrumentation, SEC and IC are essentially the same LC methods except for the choice of the stationary (column) and mobile (eluent) phase. IC and SEC separation modes can even be interconverted at the same stationary phase by changing either the eluent (composition if a mixed solvent is used) or column temperature. Figure 1 shows the effect of solvent composition and column temperature on the SEC/IC retention of polystyrene samples with narrow MWD.

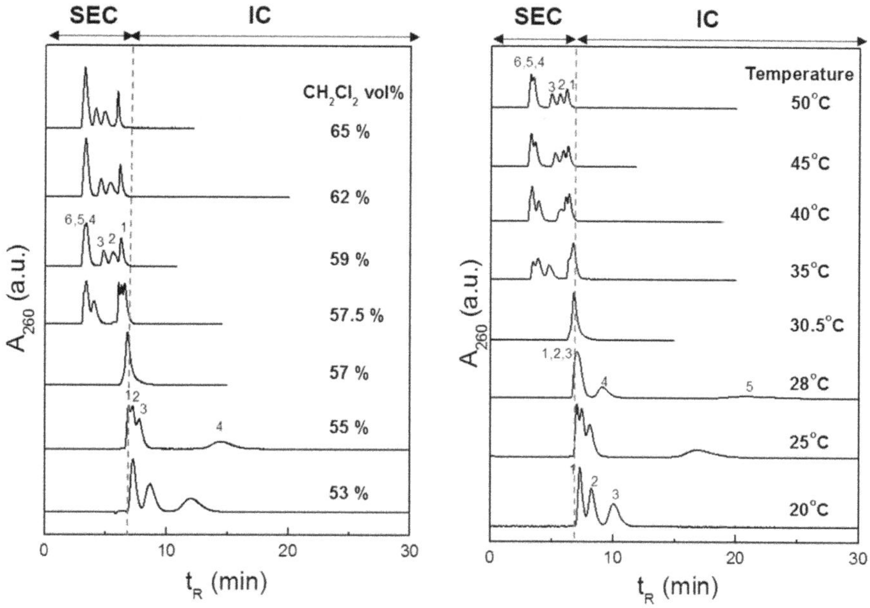

Figure 1. Chromatograms of six polystyrene standards (M_w: (1) 2.5k, (2) 12k, (3) 29k, (4) 165k, (5) 502k, (6) 1,800k). Single column (Nucleosil C18; 250 x 4.6 mm, 5 μm particle, 100 Å pore) was used and the flow rate was 0.5 mL/min. (Left) Solvent composition effect at a fixed column temperature of 30.5°C with the mixed eluents of CH_2Cl_2/CH_3CN at different compositions as labeled in the plot. (Right) Temperature effect at a fixed eluent composition of 57/43 (v/v). The isothermal chromatograms were obtained at different column temperatures as labeled in the plot. The CAP condition is established at the composition of CH_2Cl_2/CH_3CN = 57/43 (v/v) and the column temperature of 30.5°C. The vertical dashed lines indicate the elution time of the injection solvent peak. From reference (14). This figure was reproduced with permission of Wiley and Sons.

In the LC separation of polymers, there exists an interesting transition behavior from the SEC mode to the IC mode, which is called the critical adsorption point (CAP) (*15*). This behavior is the result of compensation of the exclusion and the interaction effects of polymer chains with the porous stationary phase. At this point, polymer chains undergo a 'phase transition' from the unadsorbed state to the adsorbed state (*16*). The LC method utilizing this phenomenon is called as liquid chromatography at the critical condition (LCCC). In this chapter, various LC methods used in the separation of polymers will be briefly introduced. Since SEC is a well-established method, it will not be treated further.

2. Interaction Chromatography

The IC separation of polymers began in the late 1960's using thin layer chromatography (*17, 18*). They demonstrated the successful separation of copolymers according to composition and the separation of homopolymers according to MW. Also the transition behavior between SEC and IC was observed by changing the solvent composition (*19*). Higher resolution IC separation using column LC followed later (*20, 21*). IC separation utilizes the interaction of polymeric solutes with the stationary phase and it is able to separate polymers according to their chemical composition distribution (CCD) and functionality as well as MWD. Also the band broadening in IC was found much smaller than SEC (*22, 23*). (Figure 2)

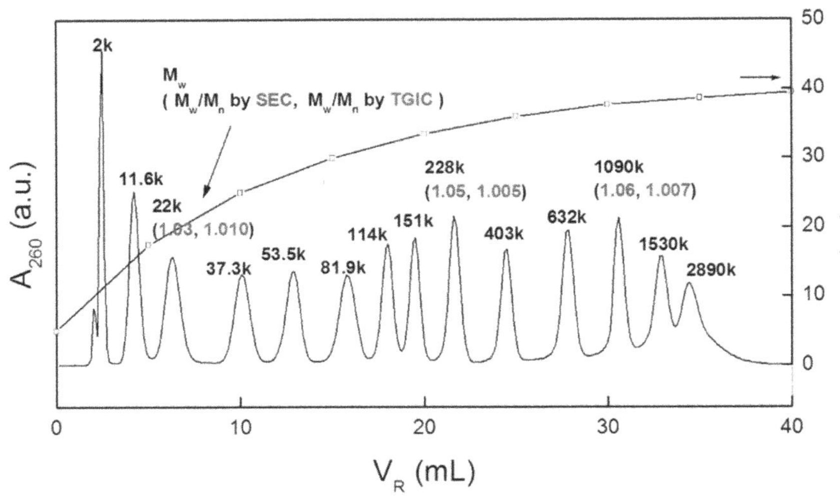

Figure 2. Temperature Gradient Interaction Chromatography (TGIC) separation of 14 standard polystyrenes showing the high resolution separation of IC. The column temperature program is shown in the plot. Mobile phase was a mixture of CH_2Cl_2/CH_3CN (57/43, v/v). Single C18 bonded silica column (Nucleosil C18, 250 x 2.1 mm, 5 μm, 100 Å) was used and the flow rate was 0.1 mL/min. TGIC shows much lower M_w/M_n values than SEC indicating higher resolution of IC than SEC. From reference (24). This figure was reproduced with permission from American Laboratory.

The sensitivity of IC to multiple molecular characteristics is a merit over SEC, however, it often makes the result of the IC separation more complicated since the retention of polymers is controlled by multiple factors. Furthermore, the IC retention of a polymer sample can vary a lot for the polymers with wide distribution in the molecular characteristics. For an example, the interaction energy of a homopolymer solute is expected to be proportional to MW (*25*) and the retention of a polymer solute (proportional to the distribution equilibrium constant) increases exponentially with its MW. Therefore, an isocratic (constant eluent composition) and isothermal IC elution of a polymer sample having a wide MWD is practically impossible. To overcome this problem, the interaction strength is controlled during the elution by changing either the solvent composition (solvent gradient elution) or column temperature (temperature gradient elution) as demonstrated in Figure 1.

In the solvent gradient interaction chromatography, solvent composition is changed from a weak to strong eluent condition so that the polymers elute in order of increasing interaction strength. This method is also called gradient polymer elution chromatography. Statistical copolymers of high MW elute near the CAP condition of corresponding chemical composition independent of MW and the solvent gradient IC is a powerful tool to separate random copolymers or polymer mixtures according to their chemical composition (*26–30*).

In temperature gradient interaction chromatography (TGIC), the eluent composition is fixed near the CAP condition, and the column temperature is programed to change the separation mode from IC to SEC (*31*, *32*) The column temperature is usually raised to promote desorption during the separation under the presumption that the adsorption of polymer is an exothermic process and the solvent strength increases with the temperature raise. But an inverse temperature dependence is also observed in a few polymer-eluent-column systems. For a successful TGIC separation of the polymers showing the inverse temperature dependency, the column temperature needs to be decreased during the elution to increase the solvent strength. A representative example is the separation of polyethylene oxide-aqueous organic solvents-C18 bonded silica columns (*33*). The inverse temperature dependence (retention increases as the column temperature increases) indicates that the adsorption process is endothermic, i.e. energetically unfavorable. Similar inverse temperature dependency was also observed in polystyrene-chloroform/n-hexane-bare silica column (*34*) and poly(methyl methacrylate)-1,4 dioxane/CH_2Cl_2-bare silica column (*35*). It shows that the adsorption of polymer chains to the stationary phase is not a simple enthalpic interaction between the polymer chains with the surface of the porous medium. Hydrophobic interaction was proposed to explain the inverse temperature dependence of polyethylene oxide with C18 bonded silica stationary phase (*36*).

While solvent gradient elution is a common HPLC technique to separate the samples of wide MWD and/or CCD, the composition change of the mobile phase limits the utility of the detectors sensitive to the composition change (e.g. refractive index, light scattering, and viscosity detectors) due to the large background signal drift. On the other hand, temperature gradient elution is an isocratic elution method that allows more freedom in choosing detectors with appropriate temperature control. However, the temperature change cannot vary the solvent strength of the eluent as much as the solvent composition variation, and TGIC is suitable for the analysis of polymer systems that need a fine control instead of a large change of the solvent strength. TGIC has been successfully applied for high resolution separation of homopolymers according to MW (Figure 2) (*22*, *23*), separation according to chemical composition (*37*), separation of polymer mixtures (Figure 3) (*38*, *39*), fractionation of individual blocks in block copolymers (*40*, *41*), separation of branched polymers (*42*, *43*), separation according to the stereoregularity (*44*), and separation according to the isotopic content (*45*, *46*).

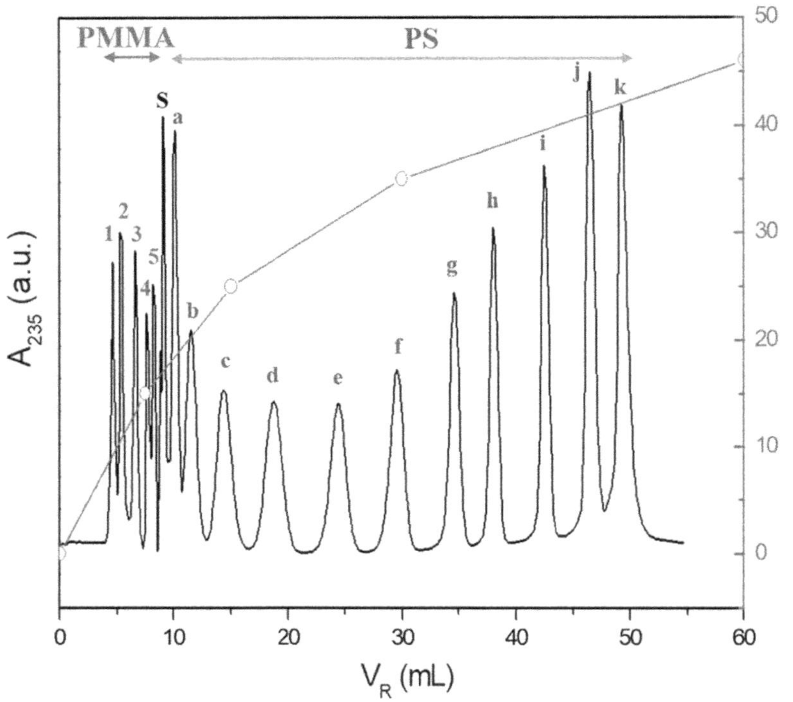

Figure 3. TGIC separation of a mixture of 10 polystyrene (a-k: 1.7k, 5.1k, 11.6k, 22.0k, 37.3k, 68.0k, 114k, 208k, 502k, 1090k, 2890k) and 5 poly(methyl methacrylate) (1-5: 1500k, 501k, 77.5k, 8.5k, 2.0k) standards. S is the injection solvent peak. The temperature program is also shown in the plot. Three columns (Nucleosil C18, 250 x 4.5 mm, 100Å & 500 Å & 1,000Å) are used and the mobile phase is a mixture of CH_2Cl_2/CH_3CN(54/46 (v/v)). From reference (38).

There are other LC techniques related to IC. One is the barrier method in which the separation mode is SEC and a narrow weak solvent plug (barrier) is injected prior to the sample injection (*47–49*). Since the polymer analytes are under SEC condition, they move faster than the barrier and catch up to the barrier in the separation column. Depending on the solvent strength of the barrier to the polymer species, some polymer species (for which the solvent strength of the plug is above the adsorption threshold) can pass the barrier while some cannot. The polymer species that cannot pass the barrier elute later with the barrier solvent plug. This method can be used effectively to separate polymer mixtures and multiple barriers of different solvent strength can be used to separate mixtures of more than two components (*50*). An evolution of the barrier method is the SEC-gradient elution. In this technique, instead of abrupt weak solvent plugs, a gradient of solvent strength (from strong to weak) is produced in the SEC column prior to the sample injection. The polymer species moves faster until it reaches the solvent composition of the adsorption threshold and then it moves together with the solvent. Polymer species can be separated depending on their adsorption threshold solvent composition (*51, 52*).

3. Critical Adsorption Point and Liquid Chromatography at Critical Condition

Critical adsorption point (CAP) is an interesting phenomenon playing an important role in the LC separation of polymers. At the CAP, the steric repulsion of the polymer segments from a solid surface is compensated by the attractive interaction with the surface. When a porous medium is used in the LC separation of polymers, the size exclusion effect of the polymers from the pores is balanced by the attractive interaction of the polymers with the surface of the packing materials (*15*, *53*, *54*), and the distribution of the polymer chains in the pores and at the interstitial space becomes the same at the CAP. The CAP condition is usually found at a separation condition (solvent composition and temperature) between the SEC mode and the IC mode elution as shown in Figure 1. A collection of the critical conditions for various polymers was reported (*55*).

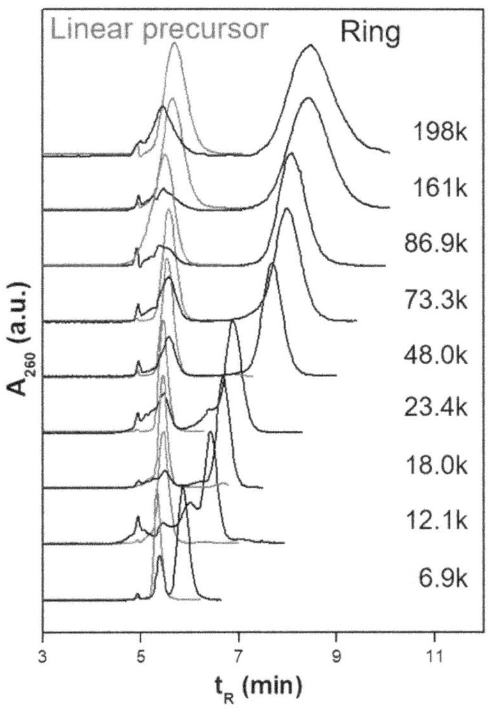

Figure 4. LCCC chromatograms of cyclic PSs and corresponding linear precursors at the CAP of linear PS. Linear precursors elute at about the same retention time (5.4 min) independent of the MW while the retention of cyclic polymer increases with molecular weight. The small peaks appearing near t_R = 5 min are the injection solvent peaks. The elution peak of cyclic polymers is completely separated from the linear precursors down to the baseline. The chromatograms of cyclic PS (black line) show that most of the samples contain contaminants, mostly the linear precursors. From reference (77).

A distinct feature observed at the CAP of a polymer species is the suppression of MW dependence in the LC retention, as the polymer elutes near the injection solvent peak independent of MW. This effect is sometimes called'chromatographic invisibility' and the 'transparency' of the polymer chain helps monitor the other molecular characteristics free from the MWD of the polymer under the critical condition. The LC technique utilizing this phenomenon is called LCCC and LCCC has been successfully employed for the separation and characterization of complex polymer systems such as the separation of individual components in polymer mixtures (*56–58*), the characterization of the individual blocks in block copolymers (*59–63*), and the separation of polymers according to the functionality (*64–71*).

Another merit of LCCC is the ability to separate polymers by chain architecture. It was found that the polymers of different architecture show CAP at different temperatures (*72*, *73*). This means that at the CAP of a linear polymer, branched polymers or cyclic polymers are retained differently and could be resolved from linear polymers. For a good example, LCCC can separate cyclic polymers from their linear precursors better than SEC or IC(Figure 4). It was theoretically predicted (*15*, *74*) and experimentally demonstrated that LCCC separation is the best available method for the purification of cyclic polymers from linear polymer byproducts present in samples synthesized by a ring-closure reaction of telechelic precursors (*75–79*). It was found that very small contamination of the linear chains changes the melt viscoelasticity of cyclic polymers dramatically (*80*). It was also found that the difference in the chain ends (the presence of different chemical groups) generated a small but noticeable effect on the CAP condition in addition to the effect of the chain architecture (*72*).

4. Two-Dimensional Liquid Chromatography (2D-LC)

In multidimensional chromatography, more than one chromatography methods are combined to separate the analytes. In practice, two-dimensional analysis using two different LC methods (2D-LC) is most widely used (*81–85*). Cross developments in thin layer chromatography is a good example of 2D-LC (*18*). In a comprehensive 2D-LC of a column chromatography, the effluent of the first dimension (1st-D) LC is injected to the second dimension (2nd-D) LC separation without loss. The chromatogram of a comprehensive 2D-LC is an overlay of the 2nd-D LC chromatograms. Therefore, repetitive 2nd-D LC separations need to be done rapidly to synchronize with the flow rate of the 1st-D LC separation and to reduce the total analysis time. Nonetheless, shortening the analysis time (achievable by high flow rate and/or by using a shorter column) may deteriorate the resolution and the analysis time and the resolution needs to be compromised for an optimum result.

Another important consideration for a successful 2D-LC separation is the solvent compatibility between the two LC separations. Since two different LC methods are combined, the injection solvent (1st-D LC solvent) is usually different from the eluent in the 2nd-D LC; poor transfer of the polymer species from the injection solvent to the eluent can deteriorate the 2nd-D LC separation. The most frequently observed problem is a break-through phenomenon in which the polymer species elute in the injection solvent plug without further separation in the 2nd-D LC (*86*). This problem occurs often when the 2nd-D LC eluent is weaker than the injection solvent (the eluent of the 1st-D LC). The compatibility problem is far less serious if the 2nd-D LC eluent is a stronger solvent for the polymer species.

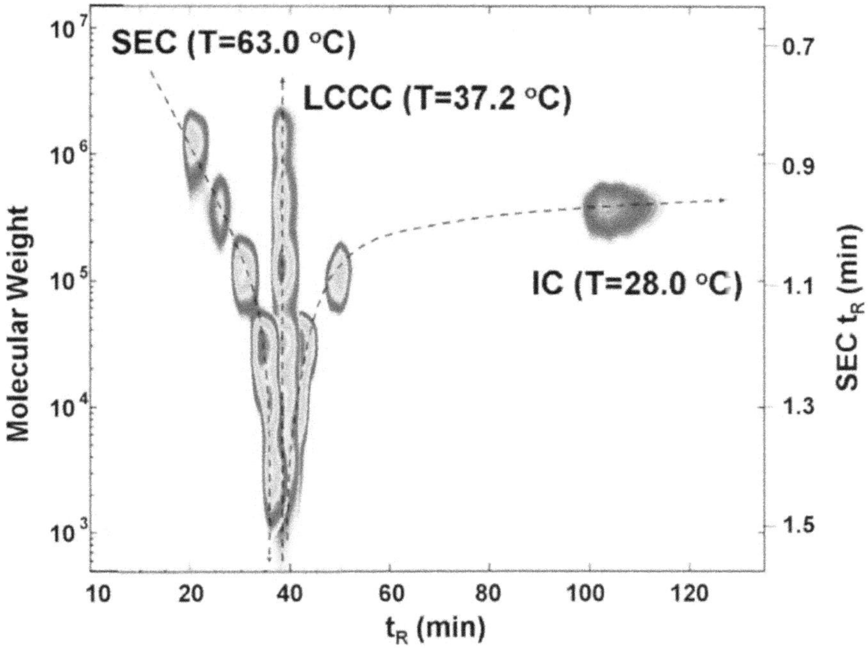

Figure 5. 2D-LC chromatograms recorded by a UV absorption detector (260 nm) showing polymer MW (SEC) vs. 1st-D LC retention time at different column temperatures; SEC (T = 63.0 °C), LCCC (T = 37.2 °C), IC (T = 28.0 °C). PS standards: 3.3k, 10k, 31k, 113k, 384k, 1800k. 1st-D IC (abscissa): Nucleosil C18 150 x 4.6 mm, 7 μm, 500 Å); Eluent: CH_2Cl_2/CH_3CN = 57/43 (v/v); Flow rate: 0.05 mL/min. 2nd-D SEC (ordinate): PL PolyPore (250 × 4.6 mm); Eluent: THF; Flow rate: 1.7 mL/min; Column temperature: 110 °C. From the reference (92). This figure was reproduced with permission from Elsevier.

The most common configuration of 2D-LC for polymer analysis is a combination of IC and SEC (IC x SEC): IC is used as the 1st-D LC to separate the polymers in terms of molecular characteristics sensitive to the IC separation and SEC is used for the 2nd-D LC for a further separation according to the chain size. SEC is suitable for a 2nd-D LC separation since the SEC eluent is a strong solvent and the solvent compatibility problem is not serious. In addition, by virtue of being an isocratic/isothermal elution, SEC separation can be repeated rapidly and multiple detection is possible. Furthermore, the SEC elution volume is always smaller than the total void volume of a column (total permeation limit for the column set), which is good for rapid/repetitive runs of the 2nd-D

LC separation. Size dependent SEC separation is often affected by molecular characteristics other than MW, but once homogeneity of the sample is improved after the 1st-D separation, SEC separation can be better correlated with MW (*43*, *87*). Furthermore, with recent developments, a reasonable SEC separation can be completed in a few minutes (*88–91*). Figure 5 shows 2D-LC (IC x SEC) contour plots of PS standards demonstrating the three different chromatographic elution modes (*92*).

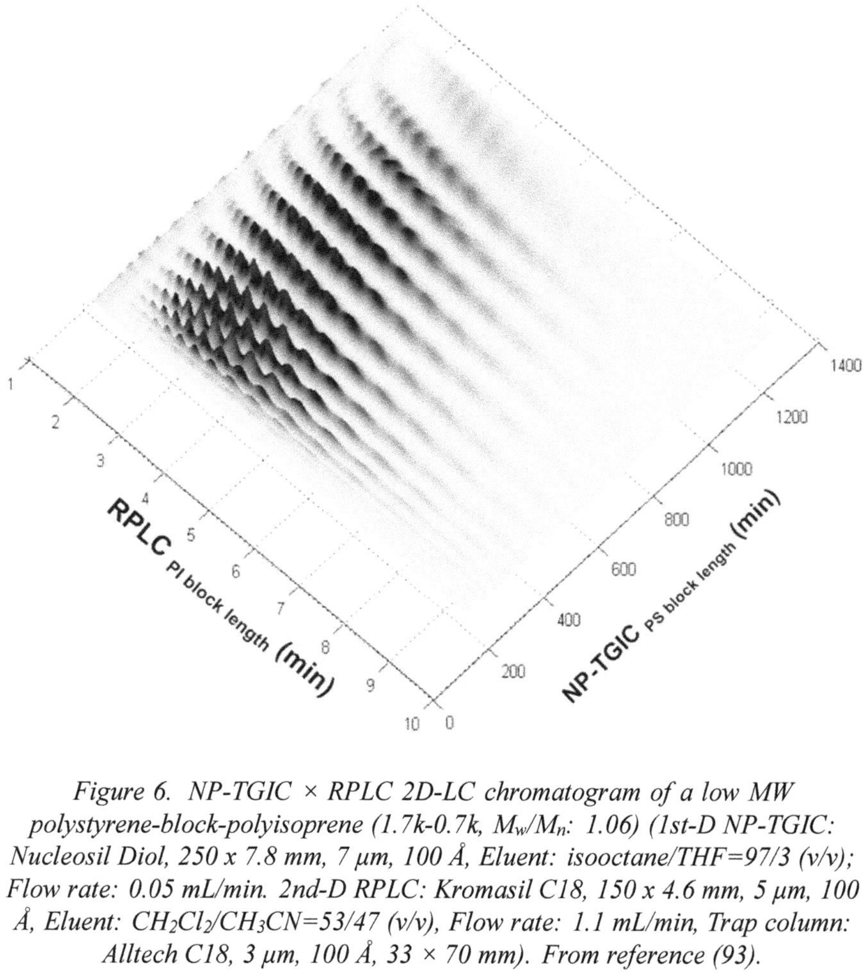

Figure 6. NP-TGIC × RPLC 2D-LC chromatogram of a low MW polystyrene-block-polyisoprene (1.7k-0.7k, M_w/M_n: 1.06) (1st-D NP-TGIC: Nucleosil Diol, 250 x 7.8 mm, 7 µm, 100 Å, Eluent: isooctane/THF=97/3 (v/v); Flow rate: 0.05 mL/min. 2nd-D RPLC: Kromasil C18, 150 x 4.6 mm, 5 µm, 100 Å, Eluent: CH_2Cl_2/CH_3CN=53/47 (v/v), Flow rate: 1.1 mL/min, Trap column: Alltech C18, 3 µm, 100 Å, 33 × 70 mm). From reference (93).

Although the use of other LC methods as a 2nd-D LC is not as popular as SEC, IC x IC and IC x LCCC have been used for more efficient characterizations of complex polymers than IC x SEC. Figure 6 shows the IC x IC 2D-LC separation of a low MW block copolymer in which a full resolution of invidual oligomers is achieved. IC as a 2nd-D LC is relatively rare since IC separation is usually not operated in an isocratic/isothermal elution mode, making it difficult for rapid/ repetitive runs. Furthermore, IC eluents are usually weak solvents and the solvent compatibility problem is more serious. Therefore, extra measures are often needed to connect two ICs without solvent compatibility problems (*46, 93*). On the other hand, LCCC can be used more easily as a 2nd-D LC since it is operated in isocratic/isothermal elution mode. If the 1st-D IC elution is done near the CAF condition, as in most TGIC separations, the solvent compatibility problem is not serious since the 1st-D IC and the 2nd-D LCCC uses a common solvent (*69, 94*)-[96].

5. Epilogue

In this brief review, the discussion is restricted toseparation with respect to the molecular characteristics of polymers, however, identification of the polymer species after the separation is important as well. Multiple detection provides with much more information for a given separation. Various methods have been developed for the detection of the HPLC effluent beyond the common concentration detectors such as UV, RI and evaporative light scattering detectors. Light scattering and viscosity detectors are no longer new detection methods but their use is still largely limited to SEC. Other useful spectroscopic methods for structural identification are infrared and nuclear magnetic resonance spectroscopic detections.[97-98] Mass spectrometry has made rapid progress in polymer analysis in recent years due to the development of soft ionization methods.[99-100] It is a powerful tool to do both separation with respect to mass and structural identification of polymers but is still limited to relatively low MW range, and various discriminations in the ionization process need to be improved to be a general tool for polymer characterization. Current use of mass spectrometry is more or less limited to the identification of narrowly dispersed fractions obtained by an adequate separation method.

Although LC separation methods are widely used for the polymer characterization at the moment, other separation methods such as hydrodynamic chromatography[101] and field flow fractionation[102-103] have their own merits where the LC methods do not work satisfactorily, e.g. very high molecular weight polymers, strongly adsorbing polymers, nanoparticles, etc. Polymers are complex mixtures and a precise characterization of the multivariate distributions is an extremely difficult, if not impossible, task. In many cases, however, the information in need can be obtained by judiciously selecting right separation and detection methods.

References

1. Moore, J. C. Gel Permeation Chromatography. I. A New Method for Molecular Weight Distribution of High Polymers. *J. Polym. Sci., Part A: Gen. Pap.* **1964**, *2*, 835–843.
2. Mori, S.; Barth, H. G. *Size Exclusion Chromatography*; Springer-Verlag: New York, 1999; 208pp.
3. Striegel, A. M.; Yau, W. W.; Kirkland, J. J.; Bly, D. D. *Modern Size-Exclusion Chromatography, Practice of Gel Permeation Chromatography and Gel Filtration Chromatography*; Wiley: Hoboken, 2009.

4. Berek, D. Size exclusion chromatography - A blessing and a curse of science and technology of synthetic polymers. *J. Sep. Sci.* **2010**, *33*, 315–335.
5. Casassa, E. F. Equilibrium distribution of flexible polymer chains between a macroscopic solution phase and small voids. *Polym. Lett* **1967**, *5*, 773–778.
6. Grubisic, Z.; Rempp, P.; Benoit, H. A Universal Calibration for Gel Permeation Chromatography. *Polym. Lett.* **1967**, *5*, 753–759.
7. Tung, L. H. Method of Calculating Molecular Weight Distribution Function from Gel Permeation Chromatograms. *J. Appl. Polym. Phys.* **1966**, *10*, 375–385.
8. Gröckner, G. *Gradient HPLC of Copolymers and Chromatographic Cross-Fractionation*; Springer -Verlag: Berlin, 1992.
9. Pasch, H. Analysis of complex polymers by interaction chromatography. *Adv. Polym. Sci.* **1997**, *128*, 1–45.
10. Berek, D. Coupled liquid chromatographic techniques for the separation of complex polymers. *Prog. Polym. Sci.* **2000**, *25*, 873–908.
11. Chang, T. Recent Advances in Liquid Chromatography Analysis of Synthetic Polymers. *Adv. Polym. Sci.* **2003**, *163*, 1–60.
12. Uliyanchenko, E.; van der Wal, S.; Schoenmakers, P. J. Challenges in polymer analysis by liquid chromatography. *Polym. Chem.* **2012**, *3*, 2313–2335.
13. Radke, W. Polymer separations by liquid interaction chromatography: Principles - prospects - limitations. *J. Chromatogr. A* **2014**, *1335*, 62–79.
14. Chang, T.; Lee, H. C.; Lee, W.; Park, S.; Ko, C. H. Polymer characterization by temperature gradient interaction chromatography. *Macromol. Chem. Phys.* **1999**, *200*, 2188–2204.
15. Gorbunov, A. A.; Skvortsov, A. M. Statistical properties of confined macromolecules. *Adv. Colloid Interface Sci.* **1995**, *62*, 31–108.
16. Skvortsov, A. M.; Gorbunov, A. A. Phase transitions in liquid chromatography of polymers. *Polym. Sci. U.S.S.R.* **1989**, *31*, 1307–1313.
17. Inagaki, H.; Matsuda, H.; Kamiyama, F. Determination of compositional heterogeneity in copolymers by thin-layer chromatography. I. Preliminary results for styrene-acrylate copolymers. *Macromolecules* **1968**, *1*, 520–5.
18. Belenkii, B. G.; Gankina, E. S. Thin-Layer Chromatography of Polymers Introductory Lecture. *J. Chromatogr.* **1970**, *53*, 3–25.
19. Belenkii, B. G.; Gankina, E. S.; Tennikov, M. B.; Vilenchik, L. Z. Fundamental Aspects of Adsorption Chromatography of Polymers and their Experimental Verification by Thin-Layer Chromatography. *J. Chromatogr.* **1978**, *147*, 99–110.
20. Teramachi, S.; Hasegawa, A.; Shima, Y.; Akatsuka, M.; Nakajima, M. Separation of Styrene-Methyl Acrylate Copolymer According to Chemical Composition, Using High-Speed Liquid Chromatography. *Macromolecules* **1979**, *12*, 992.
21. Armstrong, D. W.; Bui, K. H. Nonaqueous Reversed-Phase Liquid Chromatographic Fractionation of Polystyrene. *Anal. Chem.* **1982**, *54*, 706–708.

22. Lee, H. C.; Chang, T. Polymer Molecular Weight Characterization By Temperature Gradient High Performance Liquid Chromatography. *Polymer* **1996**, *37*, 5747–5749.
23. Lee, W.; Lee, H.; Cha, J.; Chang, T.; Hanley, K. J.; Lodge, T. P. Molecular weight distribution of polystyrene made by anionic polymerization. *Macromolecules* **2000**, *33*, 5111–5115.
24. Chang, T.; Lee, W.; Lee, H. C.; Cho, D.; Park, S. Polymer characterization by non-size exclusion chromatography: Temperature gradient interaction chromatography. *Am. Lab.* **2002**, *34*, 39.
25. Martin, A. J. P. *Biochim. Soc. Symp.* **1949**, *3*, 4.
26. Klumperman, B.; Cools, P.; Philipsen, H.; Staal, W. A qualitative study to the influence of molar mass on retention in gradient polymer elution chromatography. *Macromol. Symp.* **1996**, *110*, 1–13.
27. Berek, D. Liquid adsorption chromatography of copolymers: Molar mass (In)dependent retention. *Macromolecules* **1999**, *32*, 3671–3673.
28. Brun, Y. The mechanism of copolymer retention in interactive polymer chromatography. I. Critical point of adsorption for statistical copolymers. *J. Liq. Chromatogr. Relat. Technol.* **1999**, *22*, 3027–3065.
29. Brun, Y. The mechanism of copolymer retention in interactive polymer chromatography. II. Gradient separation. *J. Liq. Chromatogr. Relat. Technol.* **1999**, *22*, 3067–3090.
30. Bashir, M. A.; Radke, W. Predicting the chromatographic retention of polymers: Application of the polymer model to poly(styrene/ethylacrylate) copolymers. *J. Chromatogr. A* **2012**, *1225*, 107–112.
31. Ryu, J.; Chang, T. Thermodynamic prediction of polymer retention in temperature-programmed HPLC. *Anal. Chem.* **2005**, *77*, 6347–6352.
32. Chang, T. Polymer characterization by interaction chromatography. *J. Polym. Sci., Part B: Polym. Phys.* **2005**, *43*, 1591–1607.
33. Lochmüller, C. H.; Moebus, M. A.; Liu, Q. C.; Jiang, C.; Elomaa, M. Temperature Effect On Retention and Separation of Poly(Ethylene Glycol)S in Reversed-Phase Liquid Chromatography. *J. Chromatogr. Sci.* **1996**, *34*, 69–76.
34. Teramachi, S.; Matsumoto, H.; Kawai, T. Direction of temperature gradient for normal-phase temperature gradient interaction chromatography in polystyrene fractionation. *J. Chromatogr. A* **2005**, *1100*, 40–44.
35. Matsumoto, H.; Kawai, T.; Teramachi, S. Direction of temperature gradient and retention mechanism in normal-phase temperature gradient interaction chromatography for poly(methyl methacrylate) fractionation. *Polym. J.* **2007**, *39*, 458–463.
36. Cho, D.; Park, S.; Hong, J.; Chang, T. Retention mechanism of poly(ethylene oxide) in reversed-phase and normal-phase liquid chromatography. *J. Chromatogr. A* **2003**, *986*, 191–198.
37. Cong, R. J.; deGroot, W.; Parrott, A.; Yau, W.; Hazlitt, L.; Brown, R.; Miller, M.; Zhou, Z. A New Technique for Characterizing Comonomer Distribution in Polyolefins: High-Temperature Thermal Gradient Interaction Chromatography (HT-TGIC). *Macromolecules* **2011**, *44*, 3062–3072.

38. Lee, H. C.; Chang, T. Characterization of Binary Polymer Mixtures by Simultaneous Size-Exclusion Chromatography and Interaction Chromatography. *Macromolecules* **1996**, *29*, 7294–7296.
39. Park, S.; Park, I.; Chang, T.; Ryu, C. Y. Interaction-controlled HPLC for block copolymer analysis and separation. *J. Amer. Chem. Soc.* **2004**, *126*, 8906–8907.
40. Park, S.; Cho, D.; Ryu, J.; Kwon, K.; Lee, W.; Chang, T. Fractionation of block copolymers prepared by anionic polymerization into fractions exhibiting three different morphologies. *Macromolecules* **2002**, *35*, 5974–5979.
41. Chung, B. H.; Park, S.; Chang, T. HPLC fractionation and surface micellization behavior of polystyrene-b-poly(methyl methacrylate). *Macromolecules* **2005**, *38*, 6122–6127.
42. Lee, H. C.; Lee, W.; Chang, T.; Yoon, J. S.; Frater, D. J.; Mays, J. W. Linking-reaction kinetics of star shaped polystyrene by temperature gradient interaction chromatography. *Macromolecules* **1998**, *31*, 4114–4119.
43. Ahn, S.; Lee, H.; Lee, S.; Chang, T. Characterization of Branched Polymers by Comprehensive Two-Dimensional Liquid Chromatography with Triple Detection. *Macromolecules* **2012**, *45*, 3550–3556.
44. Cho, D.; Park, S.; Chang, T.; Ute, K.; Fukuda, I.; Kitayama, T. Temperature gradient interaction chromatography and MALDI-TOF mass spectrometry analysis of stereoregular poly(ethyl methacrylate)s. *Anal. Chem.* **2002**, *74*, 1928–1931.
45. Perny, S.; Allgaier, J.; Cho, D.; Lee, W.; Chang, T. Synthesis and Structural Analysis of an H-Shaped Polybutadiene. *Macromolecules* **2001**, *34*, 5408–5415.
46. Ahn, S.; Im, K.; Chang, T.; Chambon, P.; Fernyhough, C. M. 2D-LC Characterization of Comb-Shaped Polymers Using Isotope Effect. *Anal. Chem.* **2011**, *83*, 4237–4242.
47. Berek, D. Liquid chromatography of macromolecules under limiting conditions of desorption. 1. Principles of the method. *Macromolecules* **1998**, *31*, 8517–8521.
48. Rollet, M.; Pelletier, B.; Berek, D.; Maria, S.; Phan, T. N. T.; Gigmes, D. Separation of parent homopolymers from polystyrene and poly(ethylene oxide) based block copolymers by liquid chromatography under limiting conditions of desorption-3. Study of barrier efficiency according to block copolymers' chemical composition. *J. Chromatogr. A* **2016**, *1462*, 63–72.
49. Chatterjee, T.; Rickard, M. A.; Pearce, E.; Pangburn, T. O.; Li, Y. F.; Lyons, J. W.; Cong, R. J.; deGroot, A. W.; Meunier, D. M. Separating effective high density polyethylene segments from olefin block copolymers using high temperature liquid chromatography with a preloaded discrete adsorption promoting solvent barrier. *J. Chromatogr. A* **2016**, *1465*, 107–116.
50. Berek, D.; Macova, E. Liquid chromatography under limiting conditions of desorption 6: Separation of a four-component polymer blend. *J. Sep. Sci.* **2015**, *38*, 543–549.

51. Schollenberger, M.; Radke, W. SEC-Gradients, an alternative approach to polymer gradient chromatography: 1. Proof of the concept. *Polymer* **2011**, *52*, 3259–3262.
52. Maier, H.; Malz, F.; Reinhold, G.; Radke, W. SEC Gradients: An Alternative Approach to Polymer Gradient Chromatography. Separation of Poly(methyl methacrylate-stat-methacrylic acid) by Chemical Composition. *Macromolecules* **2013**, *46*, 1119–1123.
53. Skvortsov, A. M.; Gorbunov, A. A. Adsorption Effects in the Chromatogrpahy of Polymers. *J. Chromatogr.* **1986**, *358*, 77–83.
54. Ziebarth, J. D.; Wang, Y. M. Interactions of complex polymers with nanoporous substrate. *Soft Matter* **2016**, *12*, 5245–5256.
55. Macko, T.; Hunkeler, D. Liquid chromatography under critical and limiting conditions: A survey of experimental systems for synthetic polymers. *Adv. Polym. Sci.* **2003**, *163*, 61–136.
56. Pasch, H. Chromatographic investigations of macromolecules in the critical range of liquid chromatography: 3. Analysis of polymer blends. *Polymer* **1993**, *34*, 4095–4099.
57. Pasch, H.; Rode, K.; Chaumien, N. Chromatographic Investigations of Macromolecules in the Critical Range of Liquid Chromatography .9. Separation of Methacrylate-Based Polymer Blends. *Polymer* **1996**, *37*, 4079–4083.
58. Bashir, M. A.; Radke, W. Predicting the chromatographic retention of polymers: Poly(methyl methacrylate)s and polyacryate blends. *J. Chromatogr. A* **2007**, *1163*, 86–95.
59. Zimina, T. M.; Kever, J. J.; Melenevskaya, E. Y.; Fell, A. F. Analysis of Block Copolymers by High-Performance Liquid Chromatography under Critical Condition. *J. Chromatogr.* **1992**, *593*, 233–241.
60. Pasch, H.; Brinkmann, C.; Much, H.; Just, U. Chromatographic investigations of macromolecules in the "critical range" of liquid chromatography II. Two-dimensional separations of poly(ethylene oxide-block-propylene oxide). *J. Chromatogr.* **1992**, *623*, 315.
61. Pasch, H.; Trathnigg, B. *HPLC of Polymers*; Springer-Verlag: Berlin, 1997.
62. Falkenhagen, J.; Much, H.; Stauf, W.; Muller, A. H. E. Characterization of block copolymers by liquid adsorption chromatography at critical conditions. 1. Diblock copolymers. *Macromolecules* **2000**, *33*, 3687–3693.
63. Lee, W.; Cho, D.; Chang, T.; Hanley, K. J.; Lodge, T. P. (LCCC)Characterization of polystyrene-b-polyisoprene diblock copolymers by liquid chromatography at the chromatographic critical condition. *Macromolecules* **2001**, *34*, 2353–2358.
64. Belenky, B. G.; Valchikhina, M. D.; Vakhtina, I. A.; Gankina, E. S.; Tarakanov, O. G. Thin-Layer Chromatography of Oligomers. *J. Chromatogr.* **1976**, *129*, 115–124.
65. Entelis, S. G.; Evreinov, V. V.; Gorshkov, A. V. Functionality and molecular weight distribution of Telechelic polymers. *Adv. Polym. Sci.* **1986**, *76*, 129–175.
66. Gorshkov, A. V.; Much, H.; Becker, H.; Pasch, H.; Evreinov, V. V.; Entelis, S. G. Chromatographic investigations of macromolecules in the "critical range"

of liquid chromatography I. Functionality type and composition distribution in polyethylene oxide and polypropylene oxide copolymers. *J. Chromatogr. A* **1990**, *523*, 91–102.

67. Baran, K.; Laugier, S.; Cramail, H. Fractionation of functional polystyrenes, poly(ethylene oxide)s and poly(styrene)-b-poly(ethylene oxide) by liquid chromatography at the exclusion-adsorption transition point. *J. Chromatogr. B* **2001**, *753*, 139–149.
68. Jiang, X. L.; Schoenmakers, P. J.; Lou, X. W.; Lima, V.; van Dongen, J. L. J.; Brokken-Zijp, J. Separation and characterization of functional poly(n-butyl acrylate) by critical liquid chromatography. *J. Chromatogr. A* **2004**, *1055*, 123–133.
69. Im, K.; Kim, Y.; Chang, T.; Lee, K.; Choi, N. Separation of branched polystyrene by comprehensive two-dimensional liquid chromatography. *J. Chromatogr. A* **2006**, *1103*, 235–242.
70. Petit, C.; Luneau, B.; Beaudoin, E.; Gigmes, D.; Bertin, D. Liquid chromatography at the critical conditions in pure eluent: An efficient tool for the characterization of functional polystyrenes. *J. Chromatogr. A* **2007**, *1163*, 128–137.
71. Oh, J.; Kuk, J.; Lee, T.; Ye, J.; Paik, H. J.; Lee, H. W.; Chang, T. Molecular Weight Distribution of Living Chains in Polystyrene Prepared by Atom Transfer Radical Polymerization. *ACS Macro Lett.* **2017**, *6*, 758–761.
72. Ziebarth, J. D.; Gardiner, A. A.; Wang, Y. M.; Jeong, Y.; Ahn, J.; Jin, Y.; Chang, T. Comparison of Critical Adsorption Points of Ring Polymers with Linear Polymers. *Macromolecules* **2016**, *49*, 8780–8788.
73. Ahn, J.; Chang, T.; Wang, X.; Limpouchova, Z.; Prochazka, K. Influence of the Chain Architecture and the Presence of End-Groups or Branching Units Chemically Different from Repeating Structural Units on the Critical Adsorption Point in Liquid Chromatography. *Macromolecules* **2017**, *50*, 8720–8730.
74. Gorbunov, A. A.; Skvortsov, A. M. Theory of Chromatographic Separation of Linear and Ring Macromolecules. *Polym. Sci. USSR* **1987**, *29*, 1025–1031.
75. Pasch, H.; Deffieux, A.; Henze, I.; Schappacher, M.; Riquelurbet, L. Analysis of Macrocyclic Polystyrenes .1. Liquid Chromatographic Investigations. *Macromolecules* **1996**, *29*, 8776–8782.
76. Lepoittevin, B.; Dourges, M. A.; Masure, M.; Hemery, P.; Baran, K.; Cramail, H. Synthesis and characterization of ring-shaped polystyrenes. *Macromolecules* **2000**, *33*, 8218–8224.
77. Lee, H. C.; Lee, H.; Lee, W.; Chang, T.; Roovers, J. Fractionation of cyclic polystyrene from linear precursor by HPLC at the chromatographic critical condition. *Macromolecules* **2000**, *33*, 8119–8121.
78. Lee, W.; Lee, H.; Lee, H. C.; Cho, D.; Chang, T.; Gorbunov, A. A.; Roovers, J. Retention Behavior of linear and ring polystyrene at the chromatographic critical condition. *Macromolecules* **2002**, *35*, 529–538.
79. Takano, A.; Kushida, Y.; Aoki, K.; Masuoka, K.; Hayashida, K.; Cho, D.; Kawaguchi, D.; Matsushita, Y. HPLC Characterization of Cyclization

Reaction Product Obtained by End-to-End Ring Closure Reaction of a Telechelic Polystyrene. *Macromolecules* **2007**, *40*, 679–681.
80. Kapnistos, M.; Lang, M.; Vlassopoulos, D.; Pyckhout-Hintzen, W.; Richter, D.; Cho, D.; Chang, T.; Rubinstein, M. Unexpected power-law stress relaxation of entangled ring polymers. *Nat. Mater.* **2008**, *7*, 997–1002.
81. Pasch, H. Analysis of complex polymers by multidimensional techniques. . *Phys. Chem. Chem. Phys.* **1999**, *1*, 3879–3890.
82. Kilz, P. Two-dimensional chromatography as an essential means for understanding macromolecular structure. *Chromatographia* **2004**, *59*, 3–14.
83. Berek, D. Two-dimensional liquid chromatography of synthetic polymers. *Anal. Bioanal. Chem.* **2010**, *396*, 421–441.
84. Baumgaertel, A.; Altuntaş, E.; Schubert, U. S. Recent developments in the detailed characterization of polymers by multidimensional chromatography. *J. Chromatogr. A* **2012**, *1240*, 1–20.
85. Schoenmakers, P.; Aarnoutse, P. Multi-Dimensional Separations of Polymers. *Anal. Chem.* **2014**, *86*, 6172–6179.
86. Jiang, X. L.; van der Horst, A.; Schoenmakers, P. J. Breakthrough of polymers in interactive liquid chromatography. *J. Chromatogr. A* **2002**, *982*, 55–68.
87. Gerber, J.; Radke, W. Separation of linear and star-shaped polystyrenes by two-dimensional chromatography. *E-Polymers* **2005**.
88. Pasch, H.; Kilz, P. Fast liquid chromatography for high-throughput screening of polymers. *Macromol. Rapid Commun.* **2003**, *24*, 104–108.
89. Popovici, S. T.; Schoenmakers, P. J. Fast size-exclusion chromatography - Theoretical and practical considerations. *J. Chromatogr. A* **2005**, *1099*, 92–102.
90. Cho, H.; Park, S.; Ree, M.; Chang, T. Y.; Jung, J. C.; Zin, W. C. High temperature size exclusion chromatography. *Macromol. Res.* **2006**, *14*, 383–386.
91. Uliyanchenko, E.; Schoenmakers, P. J.; van der Wal, S. Fast and efficient size-based separations of polymers using ultra-high-pressure liquid chromatography. *J. Chromatogr. A* **2011**, *1218*, 1509–1518.
92. Im, K.; Park, H.-w.; Lee, S.; Chang, T. Two-dimensional liquid chromatography analysis of synthetic polymers using fast size exclusion chromatography at high column temperature. *J. Chromatogr. A* **2009**, *1216*, 4606–4610.
93. Im, K.; Park, H. W.; Kim, Y.; Chung, B. H.; Ree, M.; Chang, T. Comprehensive two-dimensional liquid chromatography analysis of a block copolymer. *Anal. Chem.* **2007**, *79*, 1067–1072
94. Murphy, R. E.; Schure, M. R.; Foley, J. P. One- and two dimensional chromatographic analysis of alcohol ethoxylates. *Anal. Chem.* **1998**, *70*, 4353–4360.

Chapter 2

Theory and Principles of Interaction Chromatography

Yongmei Wang*

Department of Chemistry, The University of Memphis, Memphis, Tennessee 38152, United States
***E-mail: ywang@memphis.edu.**

This chapter discuss theories that can help in understanding the separation of macromolecules in interaction chromatography. Beginning with an overview of the general framework of theories that can explain the separation mechanism in liquid chromatography, the author then focuses on computational studies that examine several applications of interaction chromatography to analyze block copolymers and ring polymers. The results are presented to illustrate the principles behind interaction chromatography, especially the liquid chromatography at the critical condition (LCCC).

Introduction

The terminology Interaction Chromatography (IC) used here refers collectively to a suite of polymer liquid chromatography methods operating in modes where polymer samples experience enthalpic adsorption interactions with the porous substrate in addition to entropic exclusion interactions. This is in contrast to Size Exclusion Chromatography (SEC) where polymer chains are presumed to only experience entropic exclusion from the porous substrate. IC includes Liquid Chromatography at the Critical Condition (LCCC) (*1–5*), thermal-gradient interaction chromatography (TGIC) (*6*, *7*), and gradient liquid adsorption chromatography (*8–10*). The development and applications of IC in the analysis of polymer samples containing heterogenous species have been discussed by several reviews (*7*, *11–16*). Here we will discuss theories and principles behind the IC separation in a historical context.

The starting point to understand the separation in polymer liquid chromatography is the relationship between the retention volume of polymer samples V_R with the partition coefficient K of polymer chains between the mobile phase and stationary phase, $V_R=V_0+KV_P$, where V_0 is the volume of the mobile phase and V_p is the volume of the stationary phase (i.e., the porous regions). This relationship was proposed by Casassa (*17*) under the quasichemical equilibrium assumption in the context of discussing Size Exclusion Chromatography also known as Gel Permeation Chromatography (GPC). In the case of Size Exclusion Chromatography, V_0 represents the volume of the "mobile phase" outside the porous beads of the packing materials and V_p represents the volume of the pores inside the porous beads. The partition of polymers between the two phases, characterized by partition coefficient K, determines the retention volumes of the polymers, with the implication that polymer solutes move through the column with the mobile phase outside the porous beads but are retained within the porous regions. V_0 could be determined as the exclusion limit using large solutes (polymers with high molecular weight). Solvent elution volume is related to V_0+V_p, also referred as void volume since one might assume solvent molecules can penetrate the pores with partition coefficient K=1. This same relationship however could also be extended to the Interaction Chromatography (IC), although the exact values for V_0 and V_p might vary at different chromatography conditions.

Theory for SEC

With the above relationship established, one may pursue calculation of partition coefficients K for polymers between stationary phases and mobile phases to gain an insight about retention behavior of polymers in chromatography. An important consideration for the stationary phase is the nanoscale pore confinement exerted on polymer chains while the mobile phase is considered as an unrestricted dilute polymer solution. Casassa was the first to use the Gaussian chain model for polymer chains and computed the partition coefficient K for such chains partitioning into a slit pore and a cylindrical pore. The Gaussian chain model is a mathematical treatment of polymer chain configurations, where the end-to-end vector for a polymer chain **r** is assumed to follow a Gaussian function. The distribution function $P_N(\mathbf{r})$ for a polymer chain with N-link that has end-to-end vector $\boldsymbol{r}$ satisfy the following differential equation:

$$\nabla^2 P_N(\boldsymbol{r}) - \frac{6}{b^2}\frac{\partial P_N}{\partial N} = 0 \qquad (1)$$

For a polymer chain in unconfined bulk solution, the solution to the above differential equation is a Gaussian function, $P_N(\mathbf{r}) = \left(2\sqrt{\pi}R\right)^{-3}\exp\left[-\left(\frac{\mathbf{r}}{2R}\right)^2\right]$, where $R = (Nb^2/6)^{1/2}$. The distribution function can also be written as a product of three Gaussian distribution functions for the three components, x, y and z respectively. For a chain inside a slit pore with width D, the solution needs to satisfy the boundary condition that $P_N(\mathbf{r}) = 0$ at the surface. Solutions to the above differential equation have been studied in mathematics for heat conduction problems. Therefore, one can draw on this knowledge to understand how the partition coefficient K varies with polymer molecular weight and the pore size. Important knowledge gained from these considerations is that K depends on the R/D, where R is the radius of gyration of a linear polymer chain, which is a characteristic measure of the size of polymer chains. Hence the theory supports the notion that polymer chains are separated according to size in SEC.

Theory for Interaction Chromatography (IC)

The principle of Interaction Chromatography is closely related to how a polymer chain might be adsorbed on a solid substrate. Polymer adsorption on solid substrates was an old topic in polymer science and was investigated by many approaches (*18*). It has long been recognized that when a long flexible polymer chain is brought near a surface, there is an associated loss of entropy. This loss of entropy therefore restricts the polymer chain approaching the wall. The same mechanism restricts the polymer chain entering the pore in SEC. If the polymer repeating unit has an enthalpic gain when touching the solid wall, then this enthalpic gain can compensate for the entropy loss and leads to adsorption. One characteristic associated with this transition is how the density profiles of the polymer chain near the surface varies with the distance away from the surface. When there is no enthalpic adsorption, the density profile is depleted near the wall. On the other hand, when there is enthalpic adsorption, there is an excess polymer density built-up near the surface. Let's define ε_w to denote the interaction strength of the polymer repeating unit with the wall. Then $\varepsilon_w = 0$ corresponds to the condition of SEC. However, the transition from forming a depletion layer near the wall to forming an excess density layer near the wall occurs at a critical interaction strength, $|\varepsilon_w(CC)|$ (see Figure 1 for an illustrative drawing). One may intuitively expect this critical interaction strength should depend on chain length N, as shorter chains experience less conformational entropy loss than longer chains. However, $|\varepsilon_w(CC)|$ quickly reach another critical value, commonly denoted as the Critical Adsorption Point (CAP) as chain length N increases. The CAP is normally defined as the adsorption transition point at the infinite chain length limit.

The earliest theoretical work that aims to elucidate Interaction Chromatography is by a Russian group, mainly by Skvortsov and Gorbunov (*19*, *20*). They have used the same Gaussian chain model as Casassa, but incorporated different boundary conditions. Instead of having $P_N(\mathbf{r}) = 0$ near the surface for SEC, the boundary condition changes to

$$\frac{1}{P_N}\frac{\partial P_N}{\partial z}\Big|_s = -c \qquad (2)$$

where the parameter c could be positive, negative or zero. It was recognized that CAP discussed above corresponds to the condition c=0. At the CAP, the density profile near the wall will be flat, neither depleted nor enhanced. Solutions for

$P_N(r)$ at different values of c are available, but the frequently used solution is an approximate solution valid at wide pore limit,

$$K = 1 + \frac{2R}{D}\left[\frac{(\exp((cR)^2)[1-\mathrm{erf}(-cR)]-1}{cR} - \frac{2}{\sqrt{\pi}}\right] \quad (3)$$

where the $erf(x)$ is the error function defined as $\mathrm{erf}(x) = \frac{2}{\sqrt{\pi}}\int_0^x exp(-t^2)dt$. The above solution predicts that at the CAP, K=1 regardless of the pore size D and chain length N. In the adsorption regime, however, K>1 and increases nearly exponentially with the chain length. In the SEC regime, K<1 and K decreases with chain length N. The theoretical framework provided by Skvortsov and Gorbunov has been very useful to predict and understand the Interaction Chromatography. These two investigators have worked with different experimentalists and have published many insightful joint theoretical and experimental articles (*4*, *21–23*)

Another slightly different theoretical framework that essentially produced the same results uses a lattice model of random walks to represent polymer chains. This was done by Guttman et al (*24*) using a matrix generation method. For the lattice based chain model, the value of CAP is known and is given by a simple expression, $\varepsilon_w(CAP) = \ln(Z-\lambda_1/Z)$ where Z is the lattice coordination number and λ_1 is the number of steps forbidden due to the presence of the wall and, here, the value of ε_w is reduced by the thermal energy k_BT. For a Random Walk (RW) chain on simple cubic lattice with a solid wall, ε_w (CAP) = -0.182. Both the continuum Gaussian chain model and lattice based random walk models give similar results. They both affirm that at the CAP, K=1 and is independent of chain length and pore size.

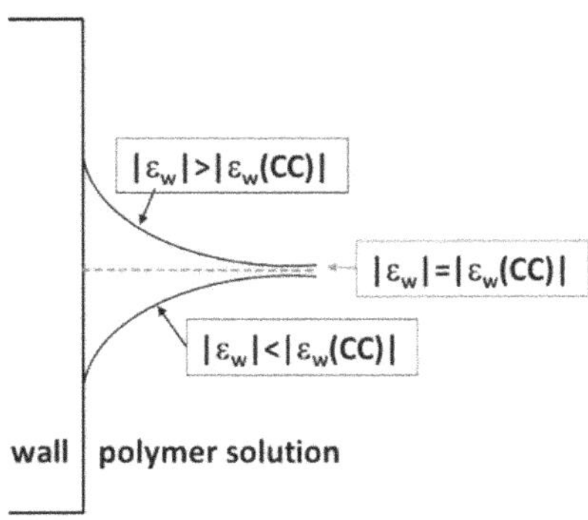

Figure 1. Illustrative drawing to show polymer density profiles near a wall at three different interaction conditions.

Computer Simulations of Interaction Chromatography

Computer simulations of polymers partitioning into pores can provide direct and insightful results, circumventing cumbersome mathematical equations that are often difficult to understand. The earliest computer simulation that examined Interaction Chromatography is by Cifra and Bleha (*25*). These two investigators used a straightforward Monte Carlo simulation approach, a lattice based simulation in a box that contained two connected regions, one region representing a confined slit pore having two interactive surfaces and the other region having periodic boundary conditions hence mimicking an unconfined bulk solution (twinbox method). Regular Monte Carlo moves governed by the Metropolis rule are used to equilibrate polymer chains between the two boxes. These authors used Self-Avoiding Walks (SAW) to represent the polymer chains, since SAW are more frequently used in computer simulations. Monte Carlo simulations provided equilibrium density profiles inside and outside the pores and the partition coefficient K is thus obtained from the average polymer densities determined. This Monte Carlo approach was first used by Wang and Teraoka (*26*) for the study of partitioning of non-dilute polymer solutions into pores. Cifra and Bleha introduced a segment/surface interaction ε_w in their simulations, and, by varying this segment/surface interaction ε_w, they investigated the transition from SEC to LAC. Perhaps because the earlier theoretical works predicted that K=1 is the condition for LCCC, these two authors tried to locate the transition point defined by K=1 condition. They found that for a fixed chain length N=100 at several pore widths D, the transition point at K=1 coincided at the same segment/surface interaction ε_w value. This crossing point was then identified as the Critical Point. However, they further showed that, at this transition point ε_w(CC), K for other chain lengths does not exactly equal one.

After hearing the interesting talk presented by Cifra about LCCC, Wang got intrigued about this transition. She and her students decided to consider this LCCC chromatography further using computer simulation. Since liquid chromatography operates at dilute polymer solution, the Monte Carlo simulation method used earlier for the study polymer partitioning into pores for non-dilute polymer solution is not a convenient approach. Hence, Wang implemented a biased chain insertion method that allowed for the determination of the excess chemical potential of a single polymer chain in the pore and in the bulk solution. Using this approach, she and her students were able to obtain the partition coefficient K at the dilute solution limit (K_0 would be a better symbol to use) for different chain lengths and slit widths at different surface/segment interaction. Wang et al (*27*) investigated the transition from SEC to LAC by using both the Random Walk (RW) model and the SAW model. The use of RW model simplifies the discussion significantly since the numerical value of CAP for RW is known precisely for RW on simple cubic lattice. Wang et al found that indeed if one uses RW model for polymer chain, K_0 is nearly independent of chain length at the CAP and is very close to one. With the SAW model, however, K_0 has a non-linear dependence on chain length. One phenomenon revealed by their simulation studies with SAW is that the transition point defined by K=1 does not occur at the same segment/surface interaction for different chain lengths. Shorter chains crossover from K<1 to K>1 at smaller segment/surface interaction $|\varepsilon_w|$ than longer chains. Nevertheless, one can still try to identify the "co-elution" point, equivalent to the experimental location of LCCC condition, as the point where K has the least dependence on chain length for a given range of chain length. Wang et al found that this co-elution point coincides to the Critical Adsorption

Points defined in the infinite chain length limit (*28*). Hence, their simulation studies using SAW model chains affirm earlier theoretical discussion that LCCC chromatography condition occurs at the CAP of a polymer chain. The only small caveat to be noted here is that *K* does not equal exactly one at the LCCC. Hence, one should not determine the theoretical critical point relevant to LCCC chromatography by looking at the point defined by *K*=1. Instead, one should try to determine the critical point following the method suggested by Wang et al (*29*), the point at which *K* has the least dependence on chain length.

Does the CAP Point Shift with Pore Size?

After establishing the link between LCCC condition and CAP, one might ask a few questions regarding the nature of CAP. One question is whether the CAP for a given polymer/surface substrate would depend on pore size? This question was discussed by several theoretical studies outside the chromatography communities (*30–33*). Wang et al investigated this problem via computer simulations for RW model and SAW model. Their earlier study in 2004 reported a shift of CAP for the SAW chains toward stronger segment/surface interaction as pore size decreases if a pore was modeled as square channel, but no shift was observed if pore was modeled as a slit pore (*29*). However, that study overlooked a small effect: polymer segments near the edge of a square channel interacts with two surface sites rather than one surface site. The observed shift in CAP as the pore size of a square channel decreases is not entirely due to the geometrical effect. Later they re-examined this problem and took care of the edge effect (*30*). They showed that for RW chain model, the CAP does not shift as the pore size of a square channel decreases after the edge effect is properly taken care of, but for SAW chain model, the CAP shifts toward stronger segment/surface interaction as the size of square channel pore decreases.

Experimental measurements of the potential shift in CAP could be performed by determining the LCCC condition for a given polymer system with chromatography columns packed with similar porous materials but with different pore sizes. Adbulahad and Ryu (*34*) performed such experiments using polystyrene standards and C18-bonded silica columns. The eluent used was fixed, a mixture of methylene chloride (MC) and acetonitrile (ACN), (MC/CAN=57/43, v/v%). The C18-bonded silica beads used have different average pore sizes (diameter =50Å, 100 Å, 120 Å and 300 Å). They determined LCCC conditions for polystyrene standards by varying the column temperature. The temperature at LCCC condition, T_{LCCC} was seen to decrease as pore size decreases, indicating a shift toward stronger segment/wall interaction. This trend of change agrees with the predictions by Wang group. However, experimental data on the dependence of T_{LCCC} on pore size could be subject to other unwanted factors, for example, the chemistry of the pore surfaces may not be exactly the same between C18-bonded silica beads with different average pore sizes. In fact, a closer examination of reported T_{LCCC} for polystyrene standards with the same fixed eluent in C18-bonded columns with different pore sizes sometimes exhibited a trend not meeting the expected change (*22*).

How Does "Chromatographically Invisible Blocks" Impact the Elution of Visible Blocks?

One interesting application of LCCC method is to mask the contribution of one block, making it chromatographically invisible, and thereby enabling the analysis of molecular weight of other blocks in block copolymers. Here "the invisible block" refer to the block that is at the LCCC condition established by its parent homopolymer. For example, if one wants to analyze the molecular weight distributions of polystyrene-*block*-polybutadiene (PS-*b*-PB) in terms of their respective blocks, the idea is that one elutes this block copolymer at the LCCC condition of polystyrene. Then the PS block is said to be the "invisible block" and the PB is the visible block. Earlier theoretical discussions suggested that partition coefficients of block copolymers should be controlled only by the visible A blocks for AB diblock, BAB triblock copolymers, star-A_mB_n with *m*-arms of A, *n*-arms of B or branched AB copolymers as long as the visible A blocks are not interposed by the invisible B blocks. Wang et al (*35–37*) examined this hypothesis carefully via computer simulations in collaboration with the experimental group led by Chang. It was found that if the visible A block is eluted in SEC mode, then the invisible B block will increase the retention of AB copolymers. The longer the invisible B block length, the larger the increase of retention. On the other hand, if the A block is eluted in the LAC mode, then the invisible B blocks will instead reduce the retention time (*38*). Figure 2 presents an example for A-graft-B copolymer where the backbone A block is set as the visible block, and grafted B blocks are the invisible blocks. As the number of "grafted" B branches increases, the influence of B blocks on A-g-B retention increases, but they exert opposite effects on the elution when A is in SEC mode versus when A is in LAC mode. These effects are observed in actual experiments. These results should also warn the experimentalists that there might be inherent errors introduced when analyzing the molecular weight of individual blocks using LCCC method. Alternative ways to estimate molecular weight of individual blocks should be employed to verify the results.

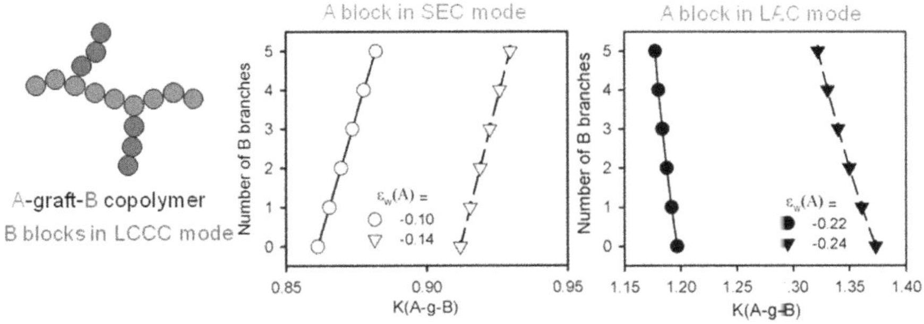

Figure 2. Impact of invisible "B" branches on the elution of A-g-B copolymer where the backbone A can be either eluted in (a) SEC mode or (b) in LAC mode. Simulation data were obtained using RW model on simple cubic lattice. Figure is reprinted from Ref (38), with permission from Elsevier, Copyright(2015).

How Does the CAP for Ring Polymers Compare with the CAP for Linear Polymers?

Ring polymers constitute another distinct class of polymers with special architecture. If one compares a polymer chain with the same number of repeating units, but one in linear chain form and the other in ring form, the number of allowed conformations of ring polymer is less than the linear chain because of the connection of the two ends. One may imagine that when ring polymer is brought near the wall, the associated loss of conformational entropy will be less than the linear chain. Hence the transition crossing over from K<1 to K>1 would occur at smaller $|\varepsilon_w(CC)|$ for a ring than for a linear chain. What this also means is that at a given chromatography condition rings are retained longer than linear chains since $K_{ring} > K_{linear}$. In fact, Chang's group in collaboration with Gorbunov has shown that rings are eluted after the linear polymers peak and is in the LAC mode if the linear polymers are eluted in LCCC mode *(22)*.

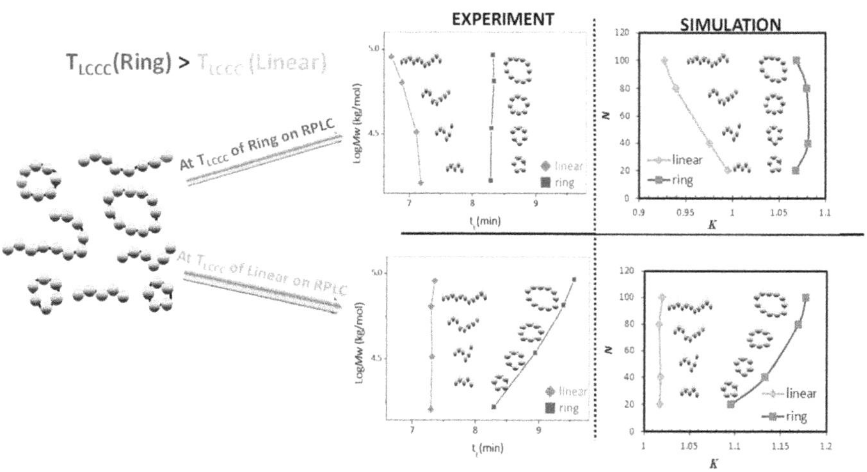

Figure 3. Comparison of retention of ring polystyrene with linear polystyrene samples in liquid chromatography column packed with C18 bonded silica surface. Figure is adapted with permission from ref (39) Copyright 2016 American Chemical Society.

If one has a set of ring polymers with different molecular weights, one could determine the coelution point of these ring polymers and denote that as the CAP for the ring polymers. How would the temperature at the coelution point for ring polymers, T_{LCCC} (ring) compares with temperature at the coelution point for the linear polymers, T_{LCCC} (linear)? This comparison can be accurately done experimentally since one could search for the T_{LCCC} for ring polymer samples and linear polymer samples on the same chromatography column and the same eluent. The only caveat that might cause complexity is the end groups in the linear chain precursor might interact differently from the rest of monomer units along the chain. Assuming there is no such complexity, how would T_{LCCC}(ring) compare with T_{LCCC} (linear)? Recently, Wang's group in collaboration with Chang's group investigated this problem both computationally and experimentally (*39*). In order to get accurate partition coefficient *K* for ring polymer, Wang group resorted back to the use of the twinbox simulation method, but implemented the Histogram Reweighting method, a relatively new Monte Carlo technique that allows for best use of statistical information obtained at different runs. They showed computationally that the co-elution point for ring polymers occurred at $|\varepsilon_w(\text{CAP, ring})|$ lower than co-elution point for linear polymers, $|\varepsilon_w(\text{CAP, linear})|$. Since in computational studies, ε_w is reduced by thermal energy. This implies at T_{LCC} (ring) should be higher than T_{LCCC}(linear). This is indeed observed in experiments. When the chromatography condition is set at T_{LCC}(ring), linear chain is eluted before ring and is in SEC mode. When the chromatography condition is set at T_{LCC}(linear), then ring polymers are eluted after the linear chains and are in LAC mode. The results are summarized in Figure 3.

Conclusion

A proper appreciation and understanding of Interaction Chromatography (IC) requires some fundamental knowledge on polymer adsorption. We have shown how knowledge gained long ago from the study of polymer adsorption on solid substrates helped the development of IC. The unique co-elution mode in LCCC is linked with the Critical Adsorption Point (CAP) known to exist in the discussion of polymer adsorption. Computer simulations that allow for flexibility to vary chain model, from RW model that does not account for excluded volume interaction to SAW model that accounts for excluded volume interaction, have been used by the author and her group to address multiple problems encountered in actual experiments. The simulations reveal interesting phenomena existed in LCCC experiments and help elucidate the true mechanism behind the observed phenomena.

Acknowledgments

The author acknowledges fruitful discussions with experimentalists in the field (Prof. Taihyun Chang, Prof. Dusan Berek, Dr. Tibor Macko, Dr. Wolfgang Radke, Dr. Y. Brun and etc.) The work discussed here have been supported in the past by the Petroleum Research Fund administered by American Chemical Society and the National Science Foundation.

References

1. Gorshkov, A. V.; Much, H.; Becker, H.; Pasch, H.; Evreinov, V. V.; Entelis, S. G. Chromatographic Investigations of Macromolecules in the "critical Range" of Liquid Chromatography. *J. Chromatogr. A* **1990**, *523*, 91–102.
2. Pasch, H.; Augenstein, M.; Trathnigg, B. Chromatographic Investigations of Macromolecules in the Critical Range of Liquid Chromatography. 6.

Analysis of the Methyl Methacrylate Block in Block Copolymers of Decyl and Methyl Methacrylate. *Macromol. Chem. Phys.* **1994**, *195*, 743–750.
3. Pasch, H.; Augenstein, M. Chromatographic Investigations of Macromolecules in the Critical Range of Liquid Chromatography. 5. Characterization of Block Copolymers of Decyl and Methyl Methacrylate. *Makromol. Chem.* **1993**, *194*, 2533–2541.
4. Skvortsov, A. M.; Gorbunov, A. A.; Berek, D.; Trathnigg, B. Liquid Chromatography of Macromolecules at the Critical Adsorption Point: Behaviour of a Polymer Chain inside Pores. *Polymer* **1998**, *39*, 423–429.
5. Berek, D.; Janco, M.; Meira, G. R. Liquid Chromatography of Macromolecules at the Critical Adsorption Point. II. Role of Column Packing: Bare Silica Gel. *J. Polym. Sci., Part A: Polym. Chem.* **1998**, *36*, 1363–1371.
6. Im, K.; Park, S.; Cho, D.; Chang, T.; Lee, K.; Choi, N. HPLC and MALDI-TOF MS Analysis of Highly Branched Polystyrene: Resolution Enhancement by Branching. *Anal. Chem.* **2004**, *76*, 2638–2642.
7. Chang, T. Polymer Characterization by Interaction Chromatography. *J. Polym. Sci., Part B: Polym. Phys.* **2005**, *43*, 1591–1607.
8. Brun, Y. The Mechanism of Copolymer Retention in Interactive Polymer Chromatography. II. Gradient Separation. *J. Liq. Chromatogr. Relat. Technol.* **1999**, *22*, 3067–3090.
9. Brun, Y.; Alden, P. Gradient Separation of Polymers at Critical Point of Adsorption. *J. Chromatogr. A* **2002**, *966*, 25–40.
10. Brun, Y. The Mechanism of Copolymer Retention in Interactive Polymer Chromatography. I. Critical Point of Adsorption for Statistical Copolymers. *J. Liq. Chromatogr. Relat. Technol.* **1999**, *22*, 3027–3065.
11. Pasch, H. Analysis of Complex Polymers by Interaction Chromatography. *Adv. Polym. Sci.* **1997**, *128*, 1–45.
12. Berek, D. Coupled Liquid Chromatographic Techniques for the Separation of Complex Polymers. *Prog. Polym. Sci.* **2000**, *25*, 873–908.
13. Chang, T. Recent Advances in Liquid Chromatography Analysis of Synthetic Polymers. *Adv. Polym. Sci.* **2003**, *163*, 1–60.
14. Trathnigg, B. Determination of Polymers of Mwd and Chemical by Chromatographic. *Prog. Polym. Sci.* **1995**, *20*, 615–650.
15. Macko, T.; Brüll, R.; Zhu, Y.; Wang, Y. A Review on the Development of Liquid Chromatography Systems for Polyolefins. *J. Sep. Sci.* **2010**, *33*, 3446–3454.
16. Radke, W. Polymer Separations by Liquid Interaction Chromatography: Principles - Prospects - Limitations. *J. Chromatogr. A* **2014**, *1335*, 62–79.
17. Casassa, E. F. Theoretical Models for Peak Migration in Gel Permeation Chromatography. *J. Phys. Chem.* **1971**, *75*, 3929–3939.
18. Fleer, G. J.; Cohen Stuart, M. A.; Scheutjens, J. M. H. M.; Cosgrove, T.; Vincent, B. *Polymers at Interfaces*; Chapman & Hall: London, 1993.
19. Skvortsov, A. M.; Gorbunov, A. A. Adsorption Effects in the Chromatography of Polymers. *J. Chromatogr.* **1986**, *358*, 77–83.
20. Gorbunov, A. A.; Skvortsov, A. M. Statistical Properties of Confined Macromolecules. *Adv. Colloid Interface Sci.* **1995**, *62*, 31–108.

21. Gorbunov, A.; Trathnigg, B. Theory of Liquid Chromatography of Mono- and Difunctional Macromolecules. *J. Chromatogr. A* **2002**, *955*, 9–17.
22. Lee, W.; Lee, H. H. C.; Cho, D.; Chang, T.; Gorbunov, A. A., Roovers, J. Retention Behavior of Linear and Ring Polystyrene at the Chromatographic Critical Condition. *Macromolecules* **2002**, *35*, 529–538.
23. Vakhrushev, A. V; Gorbunov, A. A.; Tezuka, Y.; Tsuchitani, A.; Oike, H. Liquid Chromatography of Theta-Shaped and Three-Armed Star Poly(tetrahydrofuran)s: Theory and Experimental Evidence of Topological Separation. *Anal. Chem.* **2008**, *80*, 8153–8162.
24. Guttman, C. M.; Di Marzio, E. A.; Douglas, J. F. Influence of Polymer Architecture and Polymer-Surface Interaction on the Elution Chromatography of Macromolecules through a Microporous Media. *Macromolecules* **1996**, *29*, 5723–5733.
25. Cifra, P.; Bleha, T. Steric Exclusion/adsorption Compensation in Partitioning of Polymers into Micropores in Good Solvents. *Polymer* **2000**, *41*, 1003–1009.
26. Wang, Y. M.; Teraoka, I. Computer Simulation of Semidilute Polymer Solutions in Confined Geometry : Pore as a Microscopic Probe. *Macromolecules* **1997**, *30*, 8473–8477.
27. Gong, Y.; Wang, Y. Partitioning of Polymers into Pores near the Critical Adsorption Point. *Macromolecules* **2002**, *35*, 7492–7498.
28. Ziebarth, J. D.; Wang, Y.; Polotsky, A.; Luo, M. Dependence of the Critical Adsorption Point on Surface and Sequence Disorders for Self-Avoiding Walks Interacting with a Planar Surface. *Macromolecules* **2007**, *40*, 3498–3504.
29. Orelli, S.; Jiang, W.; Wang, Y. A Computational Investigation of the Critical Condition Used in the Liquid Chromatography of Polymers. *Macromolecules* **2004**, *37*, 10073–10078.
30. Zhu, Y.; Ziebarth, J. D.; Wang, Y. Dependence of Critical Condition in Liquid Chromatography on the Pore Size of Column Substrates. *Polymer* **2011**, *52*, 3219–3225.
31. Yang, S.; Neimark, A. V. Critical Conditions of Polymer Chromatography: An Insight from SCFT Modeling. *J. Chem. Phys* **2013**, *138*, 244903.
32. Brak, R.; Owczarek, A. L.; Rechnitzer, A.; Whittington, S. G. A Directed Walk Model of a Long Chain Polymer in a Slit with Attractive Walls. *J. Phys. A: Math. Gen.* **2005**, *38*, 4309–4325.
33. van Rensburg, E. J. J.; Orlandini, E.; Owczarek, A. L.; Rechnitzer, A.; Whittington, S. G. Self-Avoiding Walks in a Slab with Attractive Walls. *J. Phys. A: Math. Gen.* **2005**, *38*, L823–L828.
34. Abdulahad, A. I.; Ryu, C. Y. Liquid Chromatography at the Critical Condition: Thermodynamic Significance and Influence of Pore Size. *J. Polym. Sci., Part B: Polym. Phys.* **2009**, *47*, 2533–2540.
35. Jiang, W.; Khan, S.; Wang, Y. Retention Behaviors of Block Copolymers in Liquid Chromatography at the Critical Condition. *Macromolecules* **2005**, *38*, 7514–7520.
36. Lee, S.; Lee, H.; Thieu, L.; Jeong, Y.; Chang, T.; Fu, C.; Zhu, Y.; Wang, Y. HPLC Characterization of Hydrogenous Polystyrene -Block- Deuterated

Polystyrene Utilizing the Isotope Effect. *Macromolecules* **2013**, *46*, 9114–9121.
37. Yang, X.; Zhu, Y.; Wang, Y. Can the Individual Block in Block Copolymer Be Made Chromatographically "Invisible" at the Critical Condition of Its Corresponding Homopolymer? *Polymer* **2013**, *54*, 3730–3736.
38. Zhu, Y.; Ziebarth, J.; Fu, C.; Wang, Y. A Monte Carlo Study on LCCC Characterization of Graft Copolymers at the Critical Condition of Side Chains. *Polymer* **2015**, *67*, 47–54.
39. Ziebarth, J. D.; Gardiner, A. A.; Wang, Y.; Jeong, Y.; Ahn, J.; Jin, Y.; Chang, T. Comparison of Critical Adsorption Points of Ring Polymers with Linear Polymers. *Macromolecules* **2016**, *49*, 8780–8788.

Chapter 3

Effects of Solvent Quality and Pore Size on a Large Scale Purification of PS-*b*-PMMA Diblock Copolymers Using Nanoporous Silica

Junwon Han[1,+] and Chang Y. Ryu[*,2]

[1]Department of Chemical and Biological Engineering, Rensselaer Polytechnic Institute, Troy, New York 12180, United States
[2]Department of Chemistry and Chemical Biology, Rensselaer Polytechnic Institute, Troy, New York 12180, United States
[*]E-mail: ryuc@rpi.edu.
[+]Current Address: Samsung Electronics, S. Korea.

The effect of solvent quality and silica nanopore size effects have been investigated for the large scale purification of polystyrene-*b*-poly(methyl methacrylate) (PS-*b*-PMMA) block copolymers. Size exclusion chromatography and solvent gradient interaction chromatography (SGIC) are employed to analyze the molecular weight distribution of the PS-*b*-PMMA block copolymers and the contents of PS homopolymer (PS_h) "contaminants" in as-synthesized block copolymers, made by anionic polymerization. In particular, for the measurements of surface access of polymers on silica surface in solution, the SGIC technique is used to quantitatively measure the concentration changes of PS_h impurities and PS-*b*-PMMA block copolymers in the supernatant solvent, before and after mixing with nanoporous silica in the mixed solvents of tetrahydrofuran (THF) and isooctane (IO). When the solvent composition of THF is at or higher than 60 % THF (v/v) in THF/IO, both PS_h and PS-*b*-PMMA do not significantly adsorb onto the nanoporous silica surface. However, the selective adsorption of PS-*b*-PMMA starts to occur when the THF composition is lowered below 55 % THF (v/v). By manipulating the THF/IO solvent compositions and choosing appropriate nanopore sized silica, it was possible to selectively adsorb about 2 grams of PS-*b*-PMMA for the purification when 10 grams of nanoporous silica is being used. Finally, as an adaptation of affinity chromatography technique, a large scale purification technique is finally demonstrated. The large scale separation technique via chromatographic filtration of the block copolymers can be easily performed to obtain multi-grams of purified block copolymers using a simple gravity column packed with silica gel. Furthermore, the chromatographic filtration technique offers the advantage of *in-situ* monitoring the loading of adsorbed PS-*b*-PMMA due to the distinct white band formation in the column.

Introduction

Macromolecular self-assembly of polystyrene-*b*-poly(methyl methacrylate) PS-*b*-PMMA block copolymers has been used as templates and building blocks for directed self-assembly in nanopatterning and nanomanufacturing (*1–5*), and various types of living polymerization techniques have been developed to enhance our ability to target the synthesis of well-defined block copolymers. Sequential living polymerization techniques are predominantly used for the synthesis of block copolymers, which usually exists as disperse mixtures in terms of chemical heterogeneity, as well as molecular weight. In particular, the sequential anionic polymerization is an attractive choice for its ability to synthesize the block copolymers with narrow molecular weight distribution. However, the preparation of well-defined PS-*b*-PMMA block copolymers via the sequential anionic polymerization is yet challenging. This is particularly due to the sensitivity of living polymeric anions towards protic impurities and oxygen. Even with rigorous purification of monomers and solvents, it is technically difficult to avoid premature terminations of the first blocks as homopolymers, presumably due to the side reactions introduced upon transferring the second monomers during the sequential anionic polymerization.

The "chromatographic filtration" technique has been developed for the complete removal of homopolymers from as-synthesized diblock copolymers, in order to overcome the synthetic difficulty of preparing homopolymers-free block copolymers (*6*, *7*). Specifically, the purification of PS-*b*-PMMA block copolymer by removing PS homopolymers (PS_h) is achieved using a simple gravity column packed with widely available silica gel for a large scale separation. Our purification strategy is that the PS-*b*-PMMA is selectively adsorbed over PS_h by controlling the solvent composition of mixed solvents using tetrahydrofuran (THF, good solvent) and isooctane (IO, poor solvent). After the un-adsorbed PS_h being "filtered" out of the silica column, the adsorbed and accumulated PS-*b*-PMMA in the column has been easily retrieved by rinsing the silica column with pure THF. In this report, the fundamental effects of solvent quality and nanopore size effects on the block copolymer adsorption will be discussed in detail, for the development of simple and large scale purification of PS-*b*-PMMA block copolymers using nanoporous silica.

Solvent gradient interaction chromatography (SGIC) provides an attractable strategy for the rapid analysis of block copolymers, because the solvent quality for the polymer adsorption can be easily manipulated by the solvent composition control using binary gradient pumps (*8*). Therefore, the SGIC offers an advantage to reduce time for multiple sample analysis. The *in-situ* mixing of good solvent and non-solvent in SGIC not only provides polymer adsorption control for separation, but also quickly "regenerates" the column at the end of SGIC for the subsequent analysis. In addition, the *in-situ* mixing of the good and poor solvents during SGIC allows us to rapidly screen the desirable solvent composition for the block copolymer purification.

Silica gel or nanoporous gravity silica column-based separation technique could be used as a simple and cost effective method to separate and purify the multi-gram scale PS-*b*-PMMA diblock copolymers. While the SGIC provide rapid and reproducible method for quantitative analysis, it is not effective for the multi-gram scale purification of block copolymers due to the fact that the SGIC still relies on the elution time-based separation that consumes a large volume of solvents. Attractive alternative to remove PS_h from PS-*b*-PMMA at a large scale is to adapt the affinity chromatography technique (*9*, *10*). The unwanted PS_h is easily removed by selectively adsorbing the wanted PS-*b*-PMMA block copolymers in a simple gravity liquid chromatography (LC) column. Taking advantage of a large specific surface area of silica gel (e.g. ~ 300 m^2/g), the economical purification of block copolymers can be achieved in multi-gram scale without generating a large amount of solvent waste. In addition, because this method is not an elution time-based separation, it can be easily implemented as a post-treatment step to purify block copolymer samples at the end of synthesis. Upon using 10 grams of nanoporous silica, with careful tuning of solvent quality and nanopore size, this report shows that it is possible to obtain about 2 grams of purified PS-*b*-PMMA block copolymers.

Experiments

Materials

Polystyrene-*b*-poly(methyl methacrylate) (PS-*b*-PMMA) diblock copolymers were either purchased from Polymer Source Inc. (Catalog # P3609-SMMA), which is made by anionic polymerization. The purchased PS-*b*-PMMA has a reported number average molecular weights of M_n = 56.8 kg/mol, M_w/M_n = 1.06, and wt_{PMMA} = 0.30. However, our in-house SEC and NMR characterization of the purchased PS-*b*-PMMA gave M_n = 68.8 kg/mol, M_w/M_n = 1.09, and wt_{PMMA} = 0.31, and we used these values for the consistency of measurements of molecular weights and composition for PS-*b*-PMMA samples before and after the purification in Results and Discussion. The PMMA composition of PS-*b*-PMMA diblock copolymer (weight fraction of PMMA, wt_{PMMA}) was characterized by

1H NMR spectroscopy (Varian, Unity-500MHz spectrometer). The nanoporous silicas with different pore sizes (Nucleosil, 7μm particle size) were purchased from *Macherey-Nagel* and were used as received. Table 1 summarizes the characteristics of different types of nucleosil silica including the average particle diameter, average pore diameter, and specific surface area.

SEC Analysis

Four SEC mixed bed columns (Shodex: KF-8025 & KF-805 & K-804, and Waters: Stryagel HR4)] were used and the eluent was tetrahydrofuran (Aldrich, HPLC grade). Chromathgrams were recorded with refractive index detector (Shodex, RI-71) and a UV detector (Acutect-500, λ= 260 nm) at a room temperature. The values of average molar masses and molar mass polydispersity index (PDI = Đ = M_w/M_n) of PS-*b*-PMMA before and after the PS homopolymers removal were measured using monodisperse PS calibration standards with a peak molecular weight ranges from 376 to 2,570,000 g/mol from Polymer Standards Service USA Inc.

SGIC Analysis

For the analysis of PS-*b*-PMMA copolymers using normal phase liquid chromatography (NPLC), a bare silica column (Zorbax Sil from Agilent with 100Å pore and 5 μm particle size, 150 mm L. × 4.6 mm I.D.) and the solvent mixture of tetrahydrofuran (THF from Fischer Scientific (HPLC grade) as a polar good solvent) and isooctane (IO from Aldrich (HPLC grade) as a non-polar poor solvent) were used as stationary and mobile phases, respectively. The HPLC apparatus consists of a dual piston pump (Lab Alliance, Series-1500), a six-port sample injector (Rheodyne, 7125) equipped with a 20 μL injection loop, and a variable UV/Vis absorption detector (Shimadzu, SPD-10Avp) operated at a wavelength of 260nm. The temperature of the column was maintained as 30°C by HPLC column temperature controller (Phenomenex, TS-130). The injection sample concentration was about 4 mg/mL, and the flow rate of the THF/IO mobile phase was fixed at 1.0 mL/min.

Adsorption of PS-b-PMMA onto Nanoporous Silica

Two milliliters of THF/IO solution of PS-*b*-PMMA of 4 mg/mL concentration was mixed with 30 mg nanoporous silica in a 20 mL vial and equilibrated for 2 hours. The supernatant solution was injected into the HPLC system. The amount of adsorbed polymer was calculated by the integral of HPLC chromatograms based on the UV absorption at the wavelength of 260 nm. The surface excess amount (Γ) of polymers was determined by the following equation.

$$\Gamma = \frac{(C_0 - C) * V}{S} \quad (1)$$

where V is the volume of polymer solution, S is the surface area of adsorbent, and C_0 and C is the concentration of original bulk solution and supernatant solution, respectively.

Purification of PS-b-PMMA in the Gravity Column

The silica gel (Aldrich, S-9258; 400-800 mesh (particle size: 15-40 μm, irregular shape), average pore size = 60 Å) was used as the packing materials for large scale purification using gravity column. As-synthesized PS-b-PMMA was dissolved in the THF/IO solvent mixture at a composition at 45/55 (v/v) and a concentration of about 10 ~ 20 mg/mL. After the polymer solution was loaded onto the LC column for the adsorption, the column was rinsed three times with THF/IO = 45/55 (v/v) to ensure the complete removal of PS homopolymers. To confirm that no polymers are eluting after the rinsing, 20 mL vial with methanol (Fisher Scientific, ACS grade) was used for checking any polymer precipitates. Finally, the gravity column was washed twice with pure THF and the desorbed polymers dissolved in the rinsing THF were finally precipitated in methanol and dried under vacuum at 50 °C for more than 24 hours.

Results and Discussion

Nanoporous Silica as Substrates for Polymer Adsorption

Nanoporous silica has been widely used as a stationary phase in liquid chromatography. Commercially available nanoporous silica particles, commonly known as nucleosil, represent a family of totally porous spherical silica materials with uniform SiO_2 structure and are typically used for high pressure liquid chromatography (HPLC) applications. The particles are uniform in size (particle diameter, D), and can have a range of pore sizes (pore diameter, D_p), typically in the rage of 50 Å (Nu 50) to 500 Å (Nu 500). Figure 1 shows the SEM surface morphology and cross-sectional TEM images of Nucleosil (Nu 300) having an average D_p of 28 nm and an average D of 7 μm.

(a)

(b)

Figure 1. (a) Scanning electron microscopy (SEM) and (b) cross-sectional transmission electron microscopy (TEM) images of nucleosil particle (Nu 300) with $D = 7\ \mu m$ and $D_p = 28.8\ nm$.

Table 1. Characteristics of Nucleosils, made by Macherey-Nagel™.

Nucleosil notation	*D (μm)*	*D_p (nm)*	*V_p (cm³/g)*	*σ (m²/g)*	*ρ (g/cm³)*
Nu 50	7.35	7.8	0.89	395	0.45
Nu 100	6.88	14.2	1.15	320	0.36
Nu 300	6.70	28.8	0.85	98	0.45
Nu 500	6.88	46.8	0.85	43	0.45

Property notation: D = average diameter of the particles; R_p = average diameter of the pores; V_p = pore volume; σ = specific surface area; ρ = density of nanoporous silica particle. These data were obtained from Nucleosil specification sheet, which was sent along with the samples.

It is important to notice that the nucleosil nanoporous particle can provide a large specific surface area in the range of 300 ~ 400 m^2 per gram. As summarized in Table 1, when the average D_p is smaller than 15 nm, the nucleosil nanoporous particle can serve as an effective substrate providing a large specific surface area higher than 300 m^2 per gram of silica. This means when about 20 grams of nucleosil are used, it is possible to provide a tremendous surface area of 6000 m^2, which is larger than American football field, that is available for the polymer adsorption and large scale purification of block copolymer. Considering that the PS-*b*-PMMA block copolymers can be adsorbed at a level of about 1 ~ 1.5 mg/m^2 on silica surface using the mixtures of THF/IO mixture, about 10 gram of nanoporous silica would be enough to obtain multi-grams of purified PS-*b*-PMMA block copolymers. However, the remaining challenge is how to selectively adsorb PS-*b*-PMMA block copolymers over PS homopolymers, and the next section in this report provides how one can control the solvent quality for the adsorption and desorption of PS-*b*-PMMA using the THF/IO solvent mixture.

Solvent Gradient Interaction Chromatography

The solvent composition plays a key role here to selectively adsorb the PS-*b*-PMMA block copolymers and to elute the PS homopolymers (PS_h) during solvent gradient interaction chromatography (SGIC). As shown in Figure 2, both PS_h and PS-*b*-PMMA are eluted at the same time, when a purely good solvent (e.g. THF) is employed is as an eluent. These profiles suggest a SEC-like retention, showing the polymer elution before the solvent. However, the 150 mm long columns are too short to provide the size-based SEC separation between PS_h and PS-*b*-PMMA. In contrast, when marginal solvents are used upon addition IO to THF, PS-*b*-PMMA is separated from PS_h via the selective adsorption of the PMMA polymers onto the silica surface. We have chosen HPLC columns with bare silica surfaces to induce the adsorption of the PMMA blocks in PS-*b*-PMMA. Figure 2 shows typical SGIC profiles of as-synthesized PS-*b*-PMMA block copolymers. For the separation of PS-*b*-PMMA from PS_h, THF and IO mixtures are employed as eluents along with silica column and the solvent composition is maintained to be marginal and isocratic at THF 55% (v/v) in THF/IO mixture for the initial three minutes after the sample injection to allow the elution of PSh. After the three minutes of isocratic elution at THF/IO = 55/45 (v/v), the solvent composition was changed at a rate of 10 % THF/min up to THF 100% for the increase of THF composition in the mobile phase. While the PS_h is eluted during the initial isocratic condition, the adsorbed PS-*b*-PMMA is eventually eluted from the column during the solvent composition gradient in SGIC.

By extending the isocratic region for the SGIC, a multiple injection technique is also applied to separate mg-scale samples for the subsequent SEC analysis of the block copolymers before and after the PS_h removal. As shown in Figure 2(b), upon 5 multiple injections under the same isocratic condition, we observed repeated elution patterns of SEC-like peaks in pairs: one from PS homopolymers and the other from injected solvents. After the elution of PS homopolymers and solvents, a larger amount of PS-*b*-PMMA was eluted than a single injection experiment due to the larger accumulation of PS-*b*-PMMA in the HPLC column.

For the large scale purification of PS-*b*-PMMA, it is important to gain fundamental understanding of the adsorption of PS_h and PS-*b*-PMMA onto the silica surface particularly as a function of THF/IO solvent compositions. We employed the SGIC method and supernatant solution approach to investigate the effect of pore size and solvent quality on the adsorption behavior of PS_h and PS-*b*-PMMA onto silica surface. When the as-synthesized PS-*b*-PMMA sample is dissolved at THF/IO = 50/50 (v/v), PS-*b*-PMMA tends to adsorb onto the silica surface. In contrast, at the same solvent composition, PS_h does not adsorb onto the silica surface. This is supported by the SGIC profiles in Figure 3(a), which shows that there is a significant reduction of PS-*b*-PMMA peak area, which is proportional to the polymer concentration in the supernatant solution, after mixing the sample solution at THF/IO = 50/50 (v/v) with nanoporous silica. In contrast, PS_h peak area remains the same after mixing with the nanoporous silica to provide evidence that the PS_h was not adsorbed onto silica surface under the solvent quality condition at THF/IO = 50/50 (v/v). When the THF composition in the sample solution was increased to THF/IO = 65/35 (v/v), this solvent quality was too good for the PS-*b*-PMMA to still remain in the supernatant solution without being adsorbed on the silica surface. Therefore, at THF/IO = 65/35 (v/v), the selective adsorption of PS-*b*-PMMA over PS_h cannot be achieved.

To elucidate the solvent quality effects on polymer adsorption at a wide range of THF/IO solvent compositions, the adsorbed amounts of polymer have been carefully and quantitatively analyzed by monitoring the polymer concentration changes in the supernatant solution before and after mixing with nanoporous silica that serves as substrates for polymer adsorption in solution. Nu 500 was chosen as the substrate, since its large pore size (D_p = 46.8 nm) allow us to measure the surface access of polymers at various THF/IO compositions without causing the complication from the size exclusion effects to interfere with the polymer adsorption. As shown in Figure 3(b), PS_h remains unadsorbed onto the silica surface at the entire range of THF/IO solvent compositions. This is supported by our observation that the concentration of PS homopolymers in the supernatant solution remains unchanged after immersing the nanoporous silica in polymer solution even at the marginal solvent quality condition at the 42/58 (v/v) THF/IO composition. On the contrary, PS-*b*-PMMA starts to adsorb onto the silica surface at the solvent composition window from 42/58 (v/v) to 55/45 (v/v) THF/IO compositions. Upon further decreasing the solvent quality by reducing THF composition in the mixed solvent, there is an increase in the surface access of PS-*b*-PMMA for having more mass of polymers to be adsorbed at the same area of silica surface. Figure 4 is provided to illustrate the selective adsorption of PS-*b*-PMMA chains over PS_h chains, when the solvent quality is decreased from 100/0 (v/v) to 45 (v/v) THF/IO. This illustration shows that PS-*b*-PMMA can be selectively adsorbed over PS_h onto the surface of nanoporous silica upon tuning the solvent quality using the mixture of THF/IO solvents.

Nanopore size effects on the adsorption behavior of PS-*b*-PMMA have been studied using a marginal solvent at THF/IO = 45/55 (v/v) composition. In particular, the surface excess dependence of polymeric species in the as-synthesized PS-*b*-PMMA block copolymers (*i.e.* PS_h vs. PS-*b*-PMMA) has been compared as a function of pore diameter (D_p). It was found that the adsorption amount of PS-*b*-PMMA is significantly increased with increasing D_p (Figure 5(a)). This is because the polymers are easily capable of accessing the pore surface inside the particles for adsorption without experiencing size exclusion effects when D_p is large. However, the higher surface excess of PS-*b*-PMMA at higher D_p would be misleading, if one tries to evaluate the effect of nanopores size on the purification capacity of PS-*b*-PMMA adsorption at a given amount (*i.e.* mass) of nanoporous particles. In other words, for the assessment of purification capacity via polymer adsorption onto nanoporous silica, it is critical to measure and understand the amount of adsorbed polymers when the same mass of nanoporous silica was used for the large scale purification. Such a polymer adsorption capacity effects on nanopores size can be best represented in Figure 5(b). This figure shows the weight ratio between the adsorbed (*i.e.* purified) PS-*b*-PMMA block copolymers and the nanoporous silica $(wt^{ad}_{PS-b-PMMA}/wt_{silica})$ at different D_p. Despite the larger surface access at higher pore diameters (See Figure 5(a)), the actual polymer adsorption capacity per unit mass of silica $(i.e.\ wt^{ad}_{PS-b-PMMA}/wt_{silica})$ is higher for the nanoporous silica with smaller D_p. This is because there is a significant reduction of specific surface area (σ in Table 1) in nanoporous silica upon the increase of D_p. When nanoporous silica is used with small pores ($D_p < 30$ nm), it is possible to adsorb about 2 grams of PS-*b*-PMMA, provided that 10 grams of nanoporous silica is being used.

Chromatographic Filtration of PS-*b*-PMMA Using Silica Gel Column

Figure 5(b) provides an important justification why it is desirable to use silica gel having a pore size of 60Å as gravity LC column materials. To reduce the cost associated for the large scale purification, irregularly-shaped silica particles (typically known as silica gel) would be useful. Furthermore, for the practical application of chromatographic filtration, the 60 Å pore size in the irregularly-shaped silica gel is an ideal choice not only for its economical availability, but also for the capacity to adsorb and purify the PS-*b*-PMMA in a large scale. However, the small pore diameter of 60 Å would make it slow to adsorb and desorb block copolymers in the inner pore surfaces for the purification. Therefore, the adsorption/desorption process time in the LC column is optimized in a time scale of 1-2 hours to ensure the thermodynamic control of the polymer adsorption/desorption by overcoming the slow transport of adsorbing polymers in nanopore confined space. Figure 6(a) shows the gravity LC column packed with silica gel for multi-gram scale purification of PS-*b*-PMMA after adding the polymer solution containing the original as-synthesized PS-*b*-PMMA sample (①). The bright white band (②) at the top of silica column corresponds to the adsorbed PS-*b*-PMMA, while the unadsorbed PS_h (③) can be washed away from the gravity LC column for the separation. As shown in Figure 6(b), the SGIC analysis on three different samples, was performed to demonstrate that "PS_h-free" PS-*b*-PMMA diblock copolymers have been retrieved by rinsing the column with THF for the desorption. In contrast, the THF/IO=45/55 (v/v) solvent rinsed solution only contains PS_h without PS-*b*-PMMA. This result confirms that PS homopolymer does not adsorb onto the nanoporous silica surface during the chromatographic filtration. The large scale purification using the gravity column offers an additional advantage in reducing time and solvent usage, when compared to HPLC. This is because it is possible to adsorb a gram of block copolymers in silica gel column without using the inject loops of finite volume (*e.g.* 100 μL) that is required for HPLC. Figure 6(c) shows the SEC profile difference before and after multi-gram purification using the gravity column. The inset SEC chromatogram of PS_h appears to be bimodal and this is because the side reactions in the anionic polymerization involves not only the premature termination, but also the coupling of the polystyrene anions. This kinds of coupling of polystyrene anions has been previous reported, when the termination involves with oxygen (*11*). Finally, as summarized in Table 2, the weight average molecular weight (Mw), polydispersity index (PDI = Đ = Mw/Mn), and PMMA volume fraction have been compared before and after purification of the as-synthesized PS-*b*-PMMA diblock copolymers. The purified PS-*b*-PMMA has narrower molecular weight distribution and higher PMMA composition, upon removal of PS homopolymers from the as-synthesized PS-*b*-PMMA.

The chromatographic filtration strategy can be applicable for the removal of homopolymers in the large-scale purification of block copolymers beyond PS-*b*-PMMA. Table 3 summarizes examples of the solvent system and temperature range with the recommended surface functional groups of silica for the selective adsorption of one block over the other block in the block copolymer purification. Depending on the average molecular weights and chemical composition of the block copolymers, the conditions of the solvent composition and temperature should be further optimized. The solvent and temperature conditions along with the types of silica particles are listed for the separation of a wide range of synthetic polymers based on adsorption liquid chromatography in review articles (*12*, *13*).

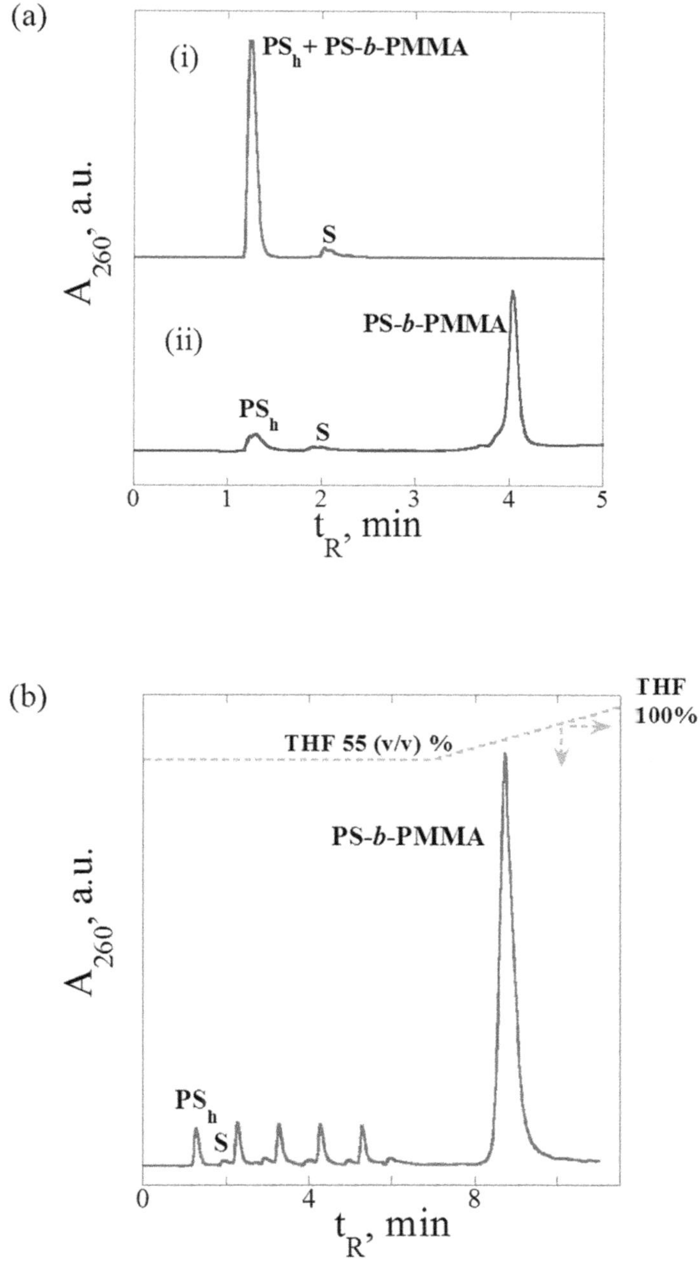

Figure 2. HPLC profiles of as-synthesized PS-b-PMMA under (a) isocratic condition using (i) neat good solvent (THF) and (ii) solvent gradient condition using mixed solvents (THF/IO) and (b) 5 multiple injections. Solvent gradient program is also shown in the plot. S = Solvent Peak.

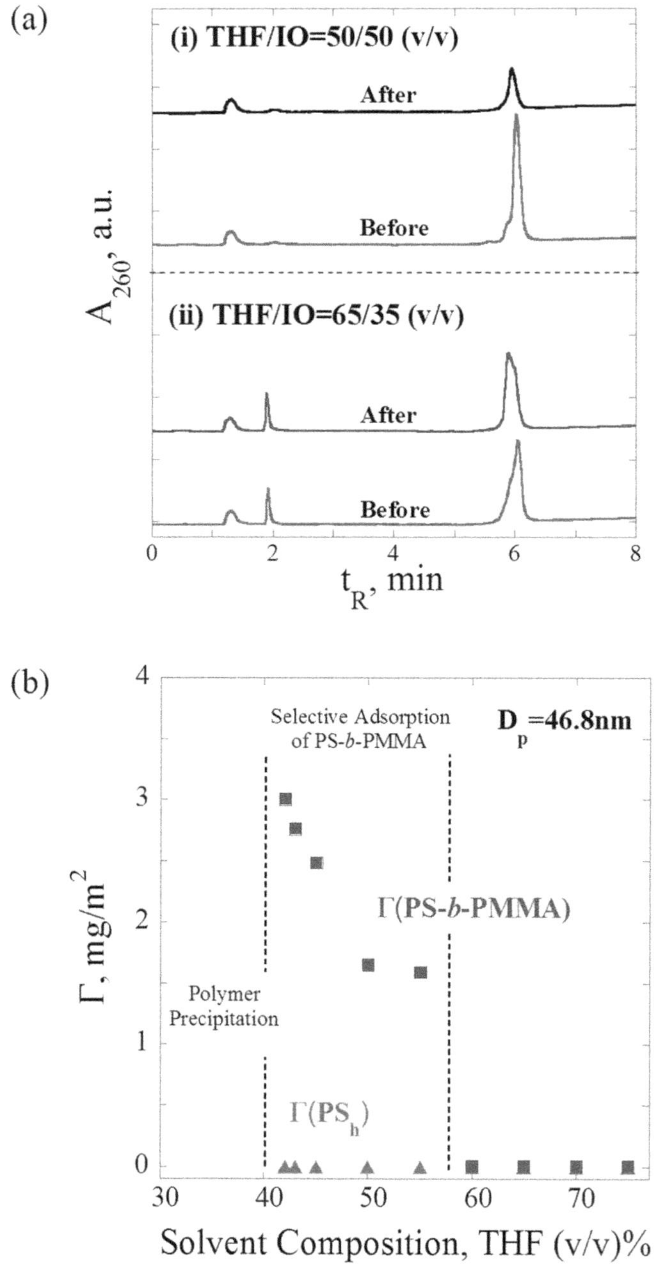

Figure 3. (a) HPLC profiles of as-synthesized PS-b-PMMA dissolved in different solvent mixture before and after the adsorption onto the nanoporous silica (Nu 500, D_p = 46.8 nm), and (b) surface excess of PS_h homopolymers and PS-b-PMMA block copolymers as a function of the mixed solvent (THF/IO) composition. Dashed line represents the precipitation point of PS-b-PMMA.

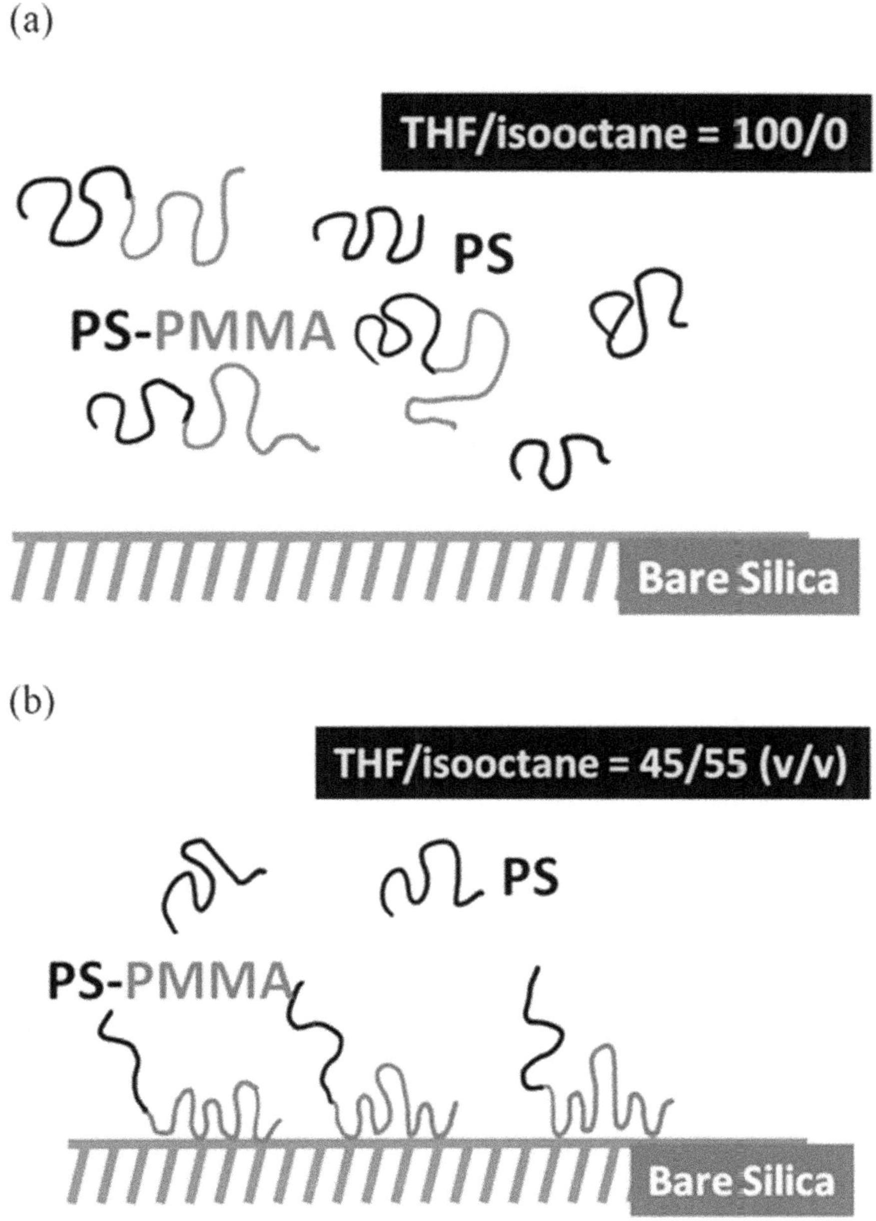

Figure 4. Schematic diagram to represent the adsorption of PS_h homopolymers and PS-b-PMMA block copolymers at different THF/IO compositions. (a) THF=IO = 100/0 (v/v) and (b) THF/IO = 45/55 (v/v)

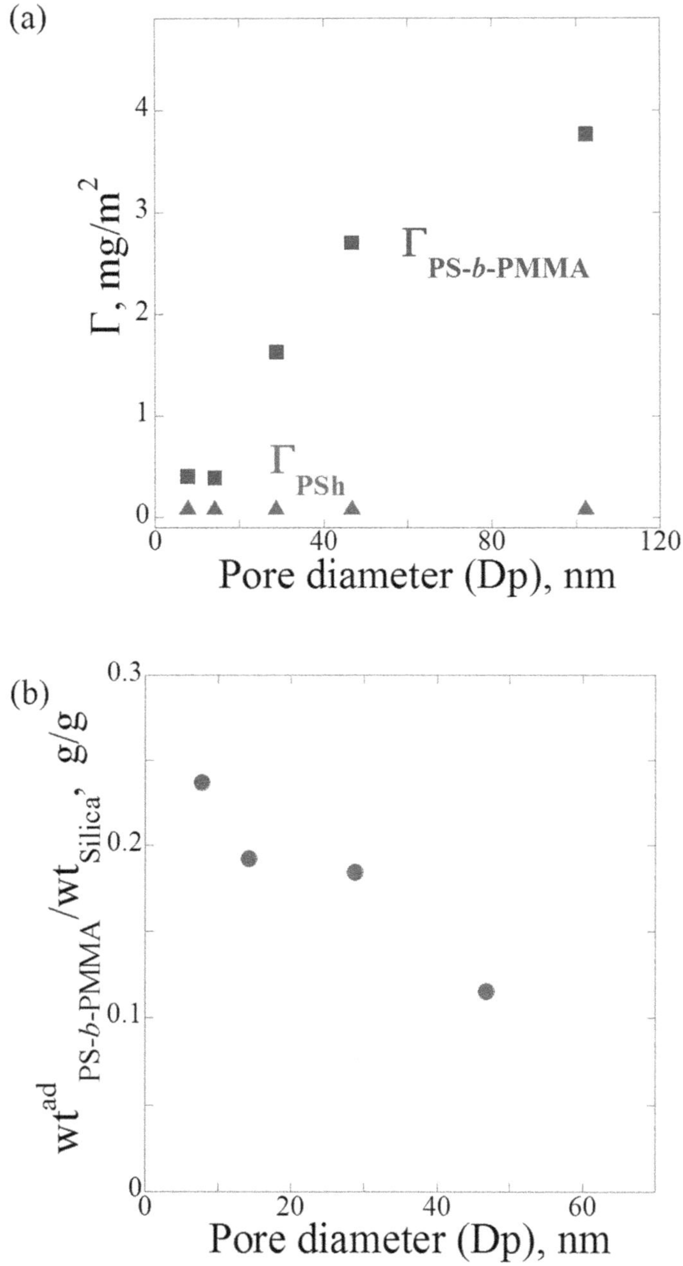

Figure 5. (a) Plot of surface excess dependence of PS_h homopolymers and PS-b-PMMA with respect to the pore diameter of nanoporous silica, and (b) weight ratio of adsorbed silica onto the silica surface to the pore diameter after dissolving as-synthesized PS-b-PMMA in 45/55 (v/v) THF/IO mixture.

Table 2. Comparison of molecular weight and polydispersity index of PS-*b*-PMMA before and after purification.

Polymer	*Before*				*After*			
	MM_w *(g/mol)*	*PDI*	wt_{PMMA}	wt_{PSh}	M_w *(g/mol)*	*PDI*	wt_{PMMA}	wt_{PSh}
PS-*b*-PMMA	68.8k	1.09	0.31	0.05	69.0k	1.02	0.34	≈ 0

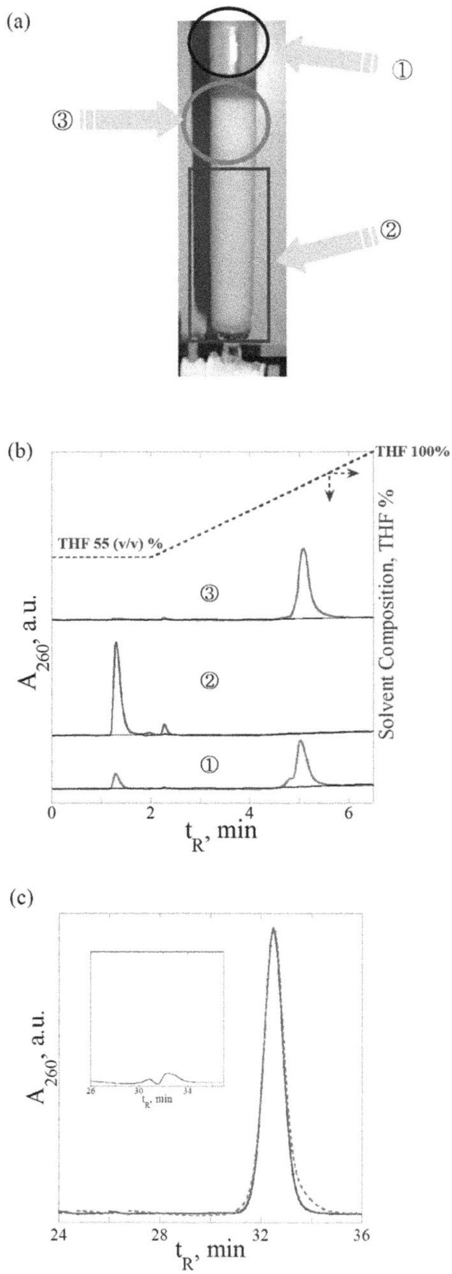

Figure 6. (a) Image of gravity LC silica gel column for selective adsorption of PS-b-PMMA, (b) SGIC profiles for the three samples indicated in (a), and (c) GPC chromatograms before (dotted line) and after (solid line) the large-scale purification via chromatographic filtration.

Table 3. Block copolymer purification conditions for the selective adsorption of one block over the other block (*i.e.* non-adsorbing block). If the contaminant is the homopolymer of the non-adsorbing block, these conditions of solvent system, temperature and silica column type can be applied for the purification of block copolymer via the chromatographic filtration.

Block Copolymers [a]	*Adsorbing Block*	*Non-Adsorbing Block (i.e., homopolymers to be removed)*	*Solvent System* [b] *(Volumetric Ratio)*	*Temperature Range*	*Surface Types of Silica Used in Column*
PS-*b*-PMMA	PMMA	PS	IO/THF = 40/60	0 – 40 °C	Bare Silica (polar)
PS-*b*-PMMA	PS	PMMA	CH_2Cl_2/ACN = 57/43	0 – 40 °C	C18-bonded Silica (non-polar)
PS-*b*-PI	PI	PS	CH_2Cl_2/ACN = 80/20	0 – 40 °C	C18-bonded Silica (non-polar)
PS-*b*-PI	PS	PI	IO/THF = 70/30	0 - 50 °C	Diol-surface Silica (polar)
PS-*b*-P2VP	P2VP	PS	THF/ACN =80/20	0 – 70 °C	Amine-surface Silica (polar)
PS-*b*-PPMA	PPMA	PS	CH_2Cl_2/ACN = 65/35	0 – 40 °C	C18-bonded Silica (non-polar)

[a] PS = polystyrene, PMMA = poly(methyl methacrylate), PI = poly(isoprene), P2VP = poly(2-vinyl pyridine), PPMA = poly(propyl methacrylate). [b] IO = isooctane, THF = tetrahydrofuran, ACN = acetonitrile, CH_2Cl_2 = methylene chloride.

Conclusions

Solvent quality and pore size of nanoporous silica plays an important role to control the amount of polymer adsorption onto the silica surface in solution. There exists a solvent composition window of THF/IO mixture between 42/58 (v/v) to 55/45 (v/v), where PS-*b*-PMMA can be selectively adsorbed over PS homopolymer (PS_h) for the removal of unwanted PS_h from the as-synthesized PS-*b*-PMMA samples. As the solvent quality is decreased by reducing the content of THF in the mixed solvent towards THF/IO = 42/58 (v/v), given that all the pore surface are available for the adsorption (i.e. using nanoporous silica with a large pore diameter (D_p) of 46.8 nm), about 3 mg of PS-*b*-PMMA can be adsorbed per m^2 of silica surface. However, any further reduction in solvent quality beyond THF/IO = 40/60 (v/v) resulted in the precipitation of polymers from the solution and thus should be avoided.

The size of nanopores also strongly influences the adsorption capacity of the nanoporous silica for the purification of PS-*b*-PMMA. As the D_p is decreased from 46.8 nm to 7.8 nm, the adsorption capacity of the nanoporous silica in terms of the weight ratio between the adsorbed (*i.e.* purified) PS-*b*-PMMA block copolymers and the nanoporous silica $(wt^{ad}_{PS-b-PMMA}/wt_{silica})$ is increased from 10% to 24% (wt/wt). This means that the multi-gram scale purification of PS-*b*-PMMA is possible when nanoporous silica particle is used for the selective adsorption substrates. Specifically, when of nanoporous silica is used, it is possible to adsorb about 2 grams of PS-*b*-PMMA for the purification, as long as the D_p is less than about 30 nm. The chromatographic filtration using silica gel with 60 Å pore has been finally applied to obtain purified PS-*b*-PMMA samples that are free from PS_h contamination. The purified PS-*b*-PMMA has a narrower polydispersity and smaller PMMA composition than the as-synthesized PS-*b*-PMMA due to removal of PS_h.

References

1. Tsai, H. Y.; Pitera, J. W.; Miyazoe, H.; Bangsaruntip, S.; Engelmann, S. U.; Liu, C. C.; Cheng, J. Y.; Bucchignano, J. J.; Klaus, D. P.; Joseph, E. A.; Sanders, D. P.; Colburn, M. E.; Guillorn, M. A. Two-Dimensional Pattern Formation Using Graphoepitaxy of PS-*b*-PMMA Block Copolymers for Advanced FinFET Device and Circuit Fabrication. *ACS Nano* **2014**, *8*, 5227.
2. Liddle, J. A.; Gallatin, G. M. Nanomanufacturing: A Perspective. *ACS Nano* **2016**, *10*, 2995.
3. Lai, F.; Borca-Tasciuc, T.; Plawsky, J. Controlling directed self-assembly of gold nanorods in patterned PS-*b*-PMMA thin films. *Nanotechnology* **2015**, *26*, 055301.
4. Jacobs, A. G.; Liedel, C.; Peng, H.; Wang, L. X.; Smilgies, D. M.; Ober, C. K.; Thompson, M. O. Kinetics of Block Copolymer Phase Segregation during Sub-millisecond Transient Thermal Annealing. *Macromolecules* **2016**, *49*, 6462.

5. Cheng, J. Y.; Rettner, C. T.; Sanders, D. P.; Kim, H. C.; Hinsberg, W. D. Dense self-assembly on sparse chemical patterns: Rectifying and multiplying lithographic patterns using block copolymers. *Adv. Mater.* **2008**, *20*, 3155.
6. Ryu, C. Y.; Han, J.; Lyoo, W. S. Block Copolymer Analysis and Purification. *J. Polym. Sci., Part B: Polym. Phys.* **2010**, *48*, 2561.
7. Kim, W.; Han, J.; Ryu, C. Y.; Yang, H. Chemical composition effects on the fracture of polystyrene-block-poly(methyl methacrylate) block copolymers. *J. Polym. Sci., Part B: Polym. Phys.* **2006**, *44*, 3612.
8. Shalliker, R. A.; Kavanagh, P. E.; Russell, I. M. Reversed-Phase Separations of High-Molecular Weight Polystyrenes in a Dichloromethane Acetonitrile Solvent System. *J. Chromatogr.* **1991**, *558*, 440.
9. Hjelm, H.; Hjelm, K.; Sjoquist, J. Protein A From Staphylococcus-Aureus - Its Isolation by Affinity Chromatography and Its Use was an Immunosorbent for Isolation of Immunoglobulins. *FEBS Lett.* **1972**, *28*, 73.
10. Cuatrecasas, P. Protein Purification by Affinity Chromatography - Derivatizations of Agarose and Polyacrylamide Beads. *J. Biol. Chem.* **1970**, *245*, 3059.
11. Park, S.; Cho, D.; Ryu, J.; Kwon, K.; Chang, T.; Park, J. Temperature gradient interaction chromatography and matrix-assisted laser desorption/ionization time-of-flight mass spectrometry analysis of air terminated polystyryllithium. *J. Chromatogr. A* **2002**, *958*, 183.
12. Chang, T. Recent Advances in Liquid Chromatography Analysis of Synthetic Polymers. *Adv. Polym. Sci.* **2003**, *163*, 1.
13. Macko, T.; Hunkler, D. Liquid Chromatography under Critical and Limiting Conditions: A Survey of Experimental Systems for Synthetic Polymers. *Adv. Polym. Sci.* **2003**, *163*, 61.

Chapter 4

Separation and Characterization of Gelatins Using Aqueous Gel Permeation Chromatography with Advanced Detection Systems

Benjamin Crawshaw,[1] Dawn Z. Herrick,[1] Wei Gao,[2] E. Peter Maziarz,[1] and X. Michael Liu*,[1]

[1]Pfizer Inc., Global Research & Development, 1211 Sherwood Avenue, Richmond, Virginia 23220, United States
[2]The Dow Chemical Company, Core R&D, Analytical Sciences, 400 Arcola Road, Collegeville, Pennsylvania 19426, United States
***E-mail: x.michael.liu@pfizer.com.Tel.: 804-257-2931.**

Gelatins are mixtures of water soluble proteins obtained by denaturation and partial hydrolysis of collagen, the main fibrous protein in animal bones, cartilage and skins. The use of gelatins plays an important role in pharmaceutical and consumer healthcare industries and have widespread applications such as hard and soft gelatin capsules, gummies, coating, and numerous drug delivery media. There is limited literature depicting the true molecular weight distributions of gelatins. This detailed information of precise and accurate molecular weights and molecular weight distributions of gelatin materials is believed to be critical for product quality and process control. In this research, we developed effective aqueous gel permeation chromatography (GPC) methods to separate and characterize five gelatin raw materials from different sources. The good mass recovery (>98%), high precision on relative molecular weight averages, M_n and M_w, (<2% standard deviations), and polydispersity of both type A and type B gelatin samples have been achieved after exploring many different GPC conditions. By coupling the GPC systems with triple detections (TD) and multi-angle light scattering detection (MALS), we utilized and compared conventional GPC, universal calibration, GPC-TD, and GPC-MALS approaches for gelatin characterization. As a result, we obtained PEO-relative molecular weights, absolute molecular weights, molecular weight distribution, and hydrodynamic and thermodynamic properties of all gelatin materials in the aqueous mobile phase that is near-physiological condition. This can be a potentially powerful tool for qualifying gelatin raw material, evaluating manufacturing processes, assessing product stabilities on the molecular level. The work here can also help troubleshoot production problems and develop new products and processes. Meanwhile, these separation and characterization methods are believed to be able to perform reverse engineering, competitive product analysis, drug-gelatin interactions, and crosslinking studies of all gelatin-based products.

Introduction

Gelatin has been widely utilized in pharmaceutical, nutraceutical, food, biochemical, and chemical industries as a gelling agent, binder, emulsifier, adhesive, and process aid. It is particularly important in pharmaceutical and nutraceutical industries as a means of active pharmaceutical ingredient protection, flavor masking, and drug delivery. Dosage forms such as hard gel capsules, soft gel capsules, and gummies are all made from gelatin. Gelatin is also used in the preparation of the active ingredients (in granulation) and as binder in tablets and beadlets. It can also be used for aqueous based film coatings.

The U.S. Pharmacopoeia / National Formulary defines gelatin as a product obtained by the partial hydrolysis of collagen derived from the skin, white connective tissue and bones of animals (*1*). Gelatin derived from an acid-treated process is known as Type A, and gelatin derived from an alkali-treated process is known as Type B. Type A is generally derived from pigskins and fish skins, while Type B is generally derived from bovine bone and hide.

In converting collagen to gelatin, the effects of acid or base hydrolysis, enzyme degradation as well as thermal stress and the time treated / age of the collagen create different types of collagen fragments. In addition to the high molecular weight fragments (proteins and/or polypeptides), there are intact and partially hydrolyzed chains. Derived from the triple helix collagen, gelatin is able to form a large variety of supramolecular structures: from globular polymer-like chains to triple-stranded helical fibrils (*1*). Hence gelatin is a mixture comprising fragments with different molecular weights and different conformations.

The certificate of analysis (CoA) of gelatin raw materials from suppliers does not provide information on molecular weight or molecular weight distribution of the material, and only lists gel strength (or Bloom strength) and solution viscosity values at a concentration of 6.67% at 60 °C. Although the solution viscosity correlates with viscosity-averaged molecular weight based on Mark-Houwink equation, this does not provide any information on the molecular weight distribution of the complex gelatin raw material. Since gelatin is known to easily crosslink as well as thermally degrade at high temperatures over time, it thus has a very broad molecular weight distribution. We believe that the detailed determination of molecular weights, molecular weight distribution and chain conformations of raw materials would help understand or predict the product performances at the molecular level. Quality control of molecular weight and molecular weight distribution of gelatins is critical for its broad applications (*2*).

The molar mass distribution of gelatins was first determined by gel filtration and ultracentrifugation fractionation and measurement by the light scattering technique (*3*). A molar mass distribution from oligo-peptides of 100 g/mol to fragments of $1x10^6$ g/mol was obtained. The molecular weight distribution determined by modern analytical methods including gel permeation chromatography (GPC) or size exclusion chromatography (SEC) has been reported to range roughly from 10,000 g/mol to 400,000 g/mol for both types of gelatin (*1*, *4*); however, molecular weights up to the millions (termed "microgel") have been reported as well (*5*). The polydispersity index (PD), which measures the broadness of a molecular weight distribution of a polymer, is the quotient of weight average molecular weight (M_w) and number average molecular weight (M_n). Gelatin normally has a broad molecular weight distribution with a PD greater than 2. The physical properties of gelatin are greatly dependent on its molecular weight and molecular weight distribution. Gel strength is mainly dependent on the population at approximately 100,000 g/mol and higher molecular weight, while the distribution at 200,000 to 400,000 g/mol has an important effect on viscosity. The viscosity of a gelatin solution correlates relatively well with the proportion of high molecular weight components (*1*).

Currently there are few SEC or GPC publications from pharmaceutical and nutraceutical industries in the scope of gelatin characterization. In this research, we developed effective aqueous GPC with triple detection (GPC-TD) and aqueous GPC with multi angle light scattering (GPC-MALS) methods to characterize type A and B gelatin raw materials under near-physiological conditions. GPC-TD provides absolute molecular weight values, intrinsic viscosity, viscometry radius and concentration at each elution moment by using a differential refractometer, a differential viscometer, and a two-angle light scattering detector. GPC-MALS yields absolute molecular weight and root mean square radius (also known as radius of gyration), and concentration at each elution moment.

Accurately and precisely characterizing raw materials has a lot of importance in pharmaceutical industrial applications. Information obtained from conventional GPC, GPC-TD and GPC-MALS allows us to: accurately characterize the molecular weight (MW), molecular weight distribution (MWD), and solution properties of gelatins. This can be a potentially powerful tool for qualifying gelatin raw material, evaluating manufacturing processes, assessing product stabilities on molecular structural level. The work here can also help us troubleshoot production problems and develop new products and processes.

Experimental

Materials

All gelatin samples were obtained from Gelita (Eberbach, Germany). The sample grade and supplied information are listed in Table 1. Narrow molecular weight poly(ethylene oxide) standards were purchased from Agilent Technologies (Santa Clara, CA, USA). Sodium chloride, sodium phosphate tribasic dodecahydrate, triethylamine (TEA) and acetonitrile were obtained from Fisher Scientific (Pittsburgh, PA, USA). Table 1 summarizes the physical properties and sources of five different gelatin sample used in our study.

Table 1. Gelatin Sample Information Provide in Certificate of Analysis

Source	*Type*	*Bloom 6.67% (g)*	*Viscosity 6.67%, 60 °C (mP)*	*Loss on Drying (%)*
Gelatin 1	B	155	33.7	11.5%
Gelatin 2	B	158	37.0	11.5%
Gelatin 3	B	162	26.0	11.6%
Gelatin 4	B	114	17.3	11.9%
Gelatin 5	A	182	29.1	11.3%

Chromatographic Conditions

Gelatin samples were prepared at 7 mg/mL using the mobile phase (5mM triethyl amine or TEA, 154mM NaCl, 20mM sodium phosphate tribasic dodecahydrate, 5% acetonitrile, pH 7.0). Dry gelatin pellets were weighed to the appropriate mass, and were allowed to swell in mobile phase at room temperature. The gelatin solutions were heated at 40°C for 16 hours to fully dissolve the gelatin. Samples were filtered with 0.45 μm nylon filter from Sartorius (Goettingen, Germany) prior to injection into the GPC. The separation of gelatin was achieved using three tandem Ultrahydrogel Linear 7.8 x 300 mm columns from Waters (Milford, MA, USA). The column set was enclosed in a column oven held at a constant temperature of 37°C. The mobile phase described above was filtered using a 0.02 μm Anodisc filter from GE Healthcare Life Sciences (Piscataway, NJ, USA) prior to use to minimize the noise level in the light scattering traces. The flow rate was set at 0.5 mL/min and sample run time was 80 minutes. One hundred microliters (100 μL) of the standard solutions or sample solutions were injected for each GPC analysis.

GPC with Triple Detection (GPC-TD)

GPC-TD Instrument and Calibration

A Malvern Viscotek GPC Max / TDA 305 instrument (Westborough, MA, USA) was used for all GPC-TD experiments. This instrument contains a pump for solvent delivery, an oven to house the columns, and three detectors: dual-angle (low angle 7° and right angle 90°) light scattering with laser wavelength of 670 nm, differential refractometer, and viscometer.

The GPC-TD system was calibrated using 3 mg/mL solution of poly(ethylene oxide) (PEO) narrow standard M_w of 48,550 g/mol (M_w/M_n = 1.06) prepared in mobile phase. The dn/dc value of the aqueous solution of PEO used for calibration was 0.1304 mL/g, as given on the certificate of analysis provided with the standard. Conventional and universal calibrations were performed with a series of PEO standards of nominal M_p molecular weights 1k, 10k, 50k, 200k, 1M g/mol. Data acquisition and calculations were performed using OmniSEC software version 5.02 from Malvern (Westborough, MA, USA).

GPC-TD Theory

As described previously (*6, 7*), the following simplified equations reflect the correlations between the detector signals and polymer molecular weight, intrinsic viscosity, and molecular radii for conformation determination:

$$R_{dRI} = K_{RI} C \frac{dn}{dc} \tag{1}$$

$$R_{visc} = K_{Visc} C[\eta] \tag{2}$$

$$R_{LS}(\theta) = K_{LS} \frac{4\pi^2 n_0^2 (\frac{dn}{dc})^2}{\lambda_0^4 N_A} CP(\theta) M_w \tag{3}$$

$$[\eta] = k_M \times M_v^a \tag{4}$$

$$R_g' = 0.328([\eta]M)^{1/3} \tag{5}$$

$$R_h = 0.251([\eta]M)^{1/3} \tag{6}$$

where R_{dRI}, R_{visc} and $R_{LS(\theta)}$ are signals of the differential refractive index detector, differential viscosity detector and light scattering detector, respectively; K_{RI}, K_{visc} and K_{LS} are the instrument constants of diffential refractive index, differential viscosity and light scattering detectors respectively. The concentration of the sample solution is denoted by C, and the specific refractive index increment is given by dn/dc. The viscosity and weight-average molecular weights are M_v and

M_w, respectively. For an individual fraction from GPC, M_v is equal to M_w since they are treated as monodisperse samples. The intrinsic viscosity of the polymer is represented by [η]. The refractive index of the solvent is denoted as n_0, λ_0 is the wavelength of the light, and N_A is the Avogadro number. The scattering angle is θ and P(θ) is the form factor, which relates the angular variation in scattering intensity to the mean square radius of the particle. The root mean square radius is also known as radius of gyration, R_g. Equation 4 is the Mark-Houwink equation, with k_M and exponent a being Mark-Houwink's constants. Equations 5 and 6 are the formulae used for calculating the radius of gyration from combination of single LS angle and viscometer (R_g'), and hydrodynamic radius (R_h) which is also more precisely known as viscometry radius and is derived from the Einstein viscosity law for particles in suspension under certain assumptions.

GPC with Multiangle Light Scattering (GPC-MALS)

GPC-MALS Instrument and Calibration

A pump, autosampler, and UV detector from Aglient 1100 series HPLC (Aglient Technologies, Santa Clara, CA) were utilized for all GPC-MALS analysis. The UV signal was monitored at the wavelength of 280 nm. An Optilab T-rEX and DAWN HELEOS II (Wyatt Technologies, Santa Barbara, CA) were used as the concentration and MALS detectors, respectively. The HELEOS II utilizes a 120 mW solid-state laser at the wavelength of 658 nm. Because a dynamic light scattering (DLS) detection module WyattQELS replaces one of the 18 MALS detectors, the MALS measures the intensity of scattered light at 17 angles. The calibration constant was calculated using toluene. The Wyatt Technology software ASTRA assumes a Rayleigh factor of 1.1705 x 10^{-5} cm^{-1} for toluene at 658 nm. Bovine serum albumin (BSA) from Sigma-Aldrich (Saint Louis, MO) at 3.3 mg/mL in mobile phase was used to determine the inter-detector delay volumes, band broadening parameters, and normalization coefficients. The dn/dc value of the aqueous solution of BSA used for calibration was 0.185 mL/g. Data acquisition and calculations were performed using ASTRA software version 6.1.4.

GPC-MALS Theory

Static light scattering is a well-known technique for dilute solution characterization. Light scattering can be used to determine the absolute molecular weight and molecular size according to equation 7:

$$\frac{Kc}{R_\theta} = \frac{1}{P(\theta)M} + 2A_2c \qquad (7)$$

where the constant K and form factor P(θ) are given as:

$$K = \frac{4\pi^2 n_0^2 (dn/dc)^2}{\lambda_0^4 N_A}$$

$$P(\theta)^{-1} = 1 + \frac{16\pi^2 \langle R_g^2 \rangle}{3\lambda_0^2} \sin^2(\theta/2)$$

and c_i is the concentration of the polymer, R_θ is the excess Rayleigh factor, θ is the scattering angle, n_0 is the refractive index of the solvent, λ_0 is the wavelength of the light, and N_A is the Avogadro number (*8*, *9*).

When light scattering is combined with GPC, Debye plots are obtained for each GPC fraction, or slice, that elutes off the column. The refractive index detector determines the concentration of each fraction, c_i. Each Debye plot can be used to determine the corresponding M_i and R_g values. By combining the concentration and molecular weight for each slice of the peak molar mass moments are determined, as defined below:

$$M_w = \frac{\sum n_i M_i^2}{\sum n_i M_i} \tag{8}$$

$$M_n = \frac{\sum n_i M_i}{\sum n_i} \tag{9}$$

where M_w and M_n represent weight-average molecular weight and number-average molecular weight of the polymer sample, respectively. The number of polymer molecules at molecular weight M_i is given by n_i. The molecular weight distribution, or polydispersity (PD), can be determined according to equation 10.

$$PD = \frac{M_w}{M_n} \tag{10}$$

Results and Discussion

Determination of Specific Refractive Index Increment (dn/dc)

The parameter dn/dc is a measure of the change in the refractive index of a polymer solution as the polymer concentration changes. It is a function of polymer structure, molar mass, solvent, light wavelength, temperature, etc. It is an essential parameter for molar mass calculation from static light scattering measurements, as demonstrated by Eq (7). Usually, it is determined by measuring the refractive index of polymer solutions at different concentrations.

Loss on drying (LOD) experiments were conducted for two days in an oven at 105°C and an average of 11.3 – 11.9% water content was obtained across all gelatin materials investigated in this study. The gelatin solutions for dn/dc measurement were prepared using raw materials without further treatment, and the sample concentrations were corrected by water content determined with the LOD experiment. The dn/dc was determined by measuring five different concentrations of a gelatin material, in the described mobile phase, using an Anton Paar Abbemet 550 refractometer at 589nm and at 37°C. The dn/dc value was determined to be 0.175 mL/g.

The dn/dc value of gelatin in water, dilute salt, phosphate buffer, at a neutral pH has been reported to range from 0.160 to 0.165 mL/g (*10–12*). The dn/dc values in our study was determined to be 0.175 mL/g, which is knowingly higher than the reported values. This can be explained by the fact that we corrected the water content of the raw materials and we used 5% of acetonitrile in mobile phase. Since the sample treatment condition was not specified by American Polymer Standards Corporation in the table on their webpage (*12*), one possible reason for this disagreement is due to the water content not being taken into account in the reference studies. A theoretical calculation of dn/dc was made using SEDFIT software (*13*, *14*). The software calculates a theoretical dn/dc value using an amino acid sequence. The amino acid composition for gelatin was found using the amino acid composition of various types of gelatin reported in the Gelatin Manufacturers Institute of America Gelatin Handbook (*15*). The software estimated a dn/dc of 0.1796 mL/g at 37°C with a 589nm laser. Remarkably this computer calculated dn/dc value is very close (Δ=0.0046) to what we determined in our experiment.

GPC Method Development

Selection of GPC Mobile Phase

Aqueous GPC is also known as gel filtration chromatography (GFC), it was originally used primarily for the separation of proteins. The interest in aqueous GPC stems from a number of sources (*16*). Many analytes, principally natural polymers, dissolve preferentially in aqueous solvents. Some biopolymers preserve their architectural properties in aqueous solvents, allowing higher-order structure to be studied by GPC. Solvent used and disposal costs are usually less when using aqueous solvents. Also, the frequent presence of non-size-exclusion effects in aqueous GPC often calls for additional studies (*17*). Non-size-exclusion effects that can potentially plague aqueous GPC analysis include ion exchange, ion inclusion, ion exclusion, intramolecular electrostatic interactions and adsorption (*18*, *19*). The selection of GPC conditions, such as column packing, mobile phases, mobile phase additives, are critical for minimizing and eliminating the non-size-exclusion effects.

In this study, different GPC conditions were evaluated and compared in terms of the injection-to-injection repeatability, mass recovery, peak shape and measured size and molecular weight at different elusion times from on-line advanced detections verse the MW calibration curve under same GPC condition. Since gelatin was made from aqueous solution and was very soluble after samples being heated to 40°C, and potential higher order structure might be able to form in aqueous solution, so all the GPC conditions studied here are aqueous phase with variables of salts, concentrations, columns, and mobile phase additives to determine the best conditions for separation and characterization of different gelatin raw materials. There have been a few published literature on characterization of gelatin and / or water soluble collagen using GPC. The GPC conditions used in the literature are summarized in Table 2.

As shown in Table 2. various salts and buffered solutions at different concentrations were used for the gelatin separation and characterization. Generally speaking, there are two different types of mobile phases, the first type (*2, 4, 20*) utilized very harsh conditions (strong surfactant 1% SDS and strong salt 1M $CaCl_2$) which completely destroyed the higher order structure and denatured the gelatin materials. Under those conditions, distinct or aggregate peaks were observed for the gelatin samples. However, these observations do not resemble the natural states of gelatin molecules in the physiological environment. The second type used more gentle salts and buffering solvents (*5, 11*). Under these conditions, the gelatin materials being analyzed were in their more natural state. In our study, we wanted to separate and characterize the gelatin materials in their natural state.

In our early stage of method development, we found $NaNO_3$ (0.1M to 0.5M), a commonly used salt in GPC, caused slight precipitation of gelatin (haziness of the solution). We also conducted the GPC experiments without any salts in the mobile phase, and observed considerable tailing, lower mass recovery, and poor injection-to-injection repeatability indicating the undesirable interactions between gelatin molecules and the column packing materials. At the end of the method development, we chose 0.9% (154mM) sodium chloride or NaCl to be the best salt for type B gelatins because it also represents physiological conditions well for the sample materials. Although gelatin in solution is amphoteric, capable of acting either as an acid or as a base, Type B gelatin has a narrow isoelectric range between pH 4.7 and 5.4 (*1*). This means that the majority of type B gelatin molecules are negatively charged at neutral pH condition (i.e. pH 7). Because of the presence of small amount of carboxyl functional groups in the column packing materials, there is minimum interaction between type B gelatin molecules and the packing materials. We can achieve satisfactory and reproducible GPC separation for Type B gelatin materials.

The molecular thermodynamic status of gelatin determined under this condition likely represents its natural existence. In addition, NaCl effectively minimizes the ionic interactions between gelatin molecules and the column packing materials. When the mobile phase is salt free, there is a much broader distribution and tailing in the light scattering chromatogram and the refractive index chromatogram has lower peak height and loss of high molecular weight population. The mass recovery without salt in the mobile phase was approximately 10% lower than that with salt. In addition, the injection-to-injection repeatability without salt in the mobile phase was found to be extremely poor. The tailing, lower mass recovery, and poor repeatability can be explained by the fact of undesirable interactions between gelatin molecules and the column packing materials.

Table 2. Summary of GPC Conditions Used in Previously Published Literature on GPC Separation and Characterization of Gelatins and/or Water Soluble Collagen

Column Set	*Column Temperature*	*Mobile Phase*	*dn/dc Reported*	*Year Published*	*Reference Source*
2 x TSK-gel PW30 + TSK-gel PW50	50°C	0.1M Na_2SO_4, 0.01M NaH_2PO_4, 1%SDS, pH 5.3	N/A	1993	(*2*)
Ultrahydrogel Linear 300x7.8mm	Ambient	0.15M NaCl, 0.1M phosphate, pH 6.8	N/A	1998	(*5*)
TSK-gel 6000 PW + TSK-gel 3000 PW	40°C,50°C, 60°C,70°C, 80°C	0.125M $LiNO_3$ or $NaNO_3$, 0.01M K phosphate, pH 6.7	0.163	2002	(*11*)
Superose 6 HR 10/30	Ambient	1M $CaCl_2$, 0.05M Tris, pH 7.5	0.190	2003	(*4*)
3 X Ultrahydrogel Linear 300x7.8mm	37°C	0.154M NaCl, 20mM Na Phosphate, 5mM TEA, 5% ACN, pH 7.0	0.175	2018	Current research

Although we were able to achieve satisfactory GPC separation for Type B gelatin materials using a mobile phase containing 0.9% NaCl, we had trouble achieving satisfactory GPC separation for Type A gelatin materials. This could be explained by the fact that the isoelectric points of Type A gelatin are much higher than those of Type B gelatins. According to the literature [1] the isoelectric points of Type A gelatins are ranging from pH 7 to 9. This means that considerable amount of Type A gelatin molecules is positively charged which can interact unfavorably with the column packing materials resulting in poor reproducibility and undesirable GPC separation. One quick solution was to add triethylamine (TEA) in the mobile phase [7]. The inclusion of 5mM of TEA can effectively reduce the undesirable column interactions. As a result, we were able to improve mass recovery, reproducibility, and achieve good GPC separation.

Under certain circumstances, it is necessary to add small amount of organic solvent/modifiers into the mobile phase to minimize the hydrophobic interactions between sample molecules and the column packing material. Herman and Field demonstrated the 100% recovery of poly(vinyl pyrrolidone) could be achieved by using 40% acetonitrile in 0.01M KH_2PO_4, pH 2.1, while the recovery could be as low as 0–25% when using water as elute (*21*). Furthermore, the addition of acetonitrile can effectively inhibit the bacterial growth of the salt solution. In our study, 5% of acetonitrile was added in the mobile phase to inhibit bacterial growth and prevent hydrophobic interactions between gelatin molecules and column packing materials. The addition of acetonitrile makes the column packing materials in a tighter swelling state, which elutes the sample molecules slightly faster than that of no acetonitrile in the mobile phase. The broader distribution of the sample with the acetonitrile addition suggests that the acetonitrile addition provides better separation. The normal MWD curve with minimal tailing with the acetonitrile addition suggests that the small amount of acetonitrile has minimized the hydrophobic interactions between the gelatin molecules and the column materials. In addition, we used 20 mM sodium phosphate to buffer the pH of the mobile phase to pH 7.0 to keep the gelatin molecules in their most natural state. Consequently we used 0.9% NaCl, 20 mM Na_2HPO_4 at pH 7.0 and 5 mM of TEA with 5% acetonitrile as the mobile phase to achieve optimal size exclusion separation with minimal ionic and hydrophobic interactions.

Selection of GPC Column Set

After we finalized our mobile phase selection of 0.9% NaCl, 20 mM Na_2HPO_4 (pH 7.0) and 5mM TEA and 5% acetonitrile, we started to evaluate different GPC column sets. In the early stage of our GPC column set selection, we tried to use a combination of two or three GPC columns with various single pore size columns including Shodex OHpak SB 806, 805, 804, 803 and 802.5HQ (Showa Denko, K.K., Tokyo, Japan). The chromatograms acquired were poor in peak shapes and considerable tailings in the low molecular weight regions These peak distortions were probably caused by the extremely broad polydispersity of gelatin materials. Since each single pore size column focused on specific hydrodynamic volume region of the sample, it is unpractical to use more than 4 GPC columns for the analysis in terms of time and cost for the sample analysis and future methods transfer to the quality control laboratories.

Based on the assumption of super broad polydispersity of gelatin materials, we decided to try the mixed-bed linear columns. At first we used 2 Waters Ultrahydrogel linear (Waters Corporation, Milford, MA, U.S.A.). Reasonable separation for most gelatin samples was obtained. However there was slight interference between the solvent peak and the low end of the lowest molecular weight gelatin sample. After adding one more Ultrahydrogel linear column, we were able to achieve satisfactory separation of high molecular weight end and free from the solvent peak interference. As a result, we settled on 3 Ultrahydrogel linear columns for our GPC separation of gelatin materials for this study. The column temperature was set at 37° C to mimic the body temperature at which the gelatin molecules are in their nearly natural state.

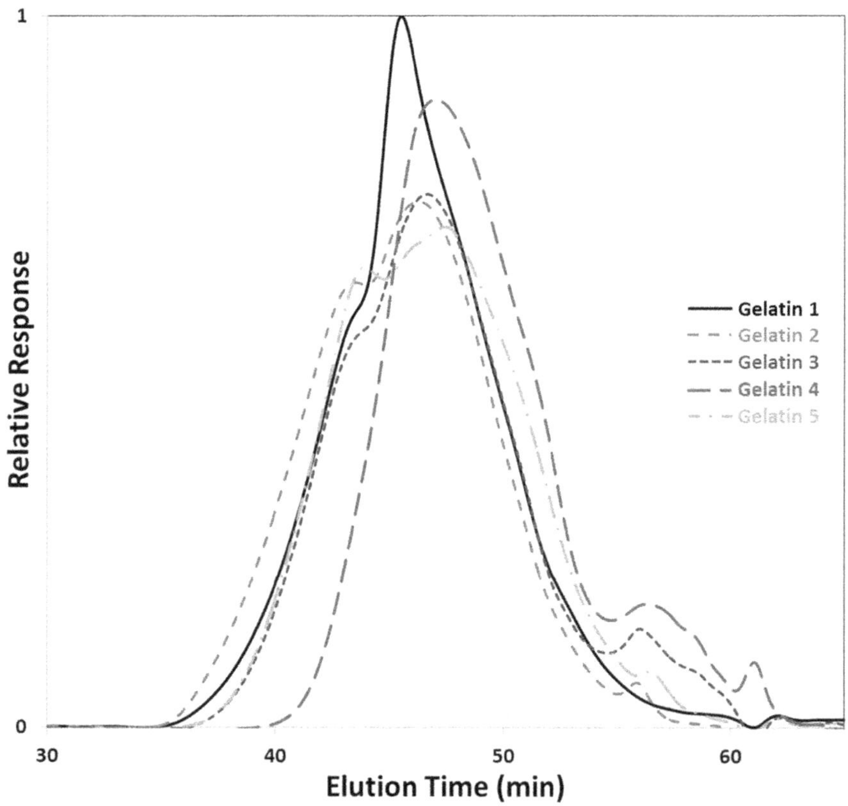

Figure 1. GPC chromatograms of five different gelatin materials. Y-axis is relative dRI detector response. The column set selected was 3X Waters Ultrahydrogel Linear and the mobile phase used was 154mM NaCl, 5mM TEA, 20mM sodium phosphate tribasic dodecahydrate, 5% acetonitrile adjusted at pH 7.0.

After the investigation of the mass recovery, peak shape, and repeatability of multiple injections, the best GPC condition used in all following studies were Ultrahydrogel Linear in a mobile phase of 5mM TEA, 154mM NaCl, 20mM sodium phosphate tribasic dodecahydrate, 5% acetonitrile, pH 7.0. The GPC chromatograms of the 5 samples were overlaid in Figure 1. The five samples analyzed showed very different and broad molecular weight profiles with bi-modal (Gelatin 1) and / or multi-modal distributions. Gelatin 2 has highest molecular weights while Gelatin 4 has the lowest molecular weights. Gelatin samples 3 and 4 have distinct late eluting peaks which may suggest the presence of low molecular weight components.

Determination of Molecular Weight (MW) and Thermodynamic Status of Gelatins in GPC Mobile Phase

Conventional Column Calibration Approach for MW and MWD

It has been well accepted that a GPC elution curve can be viewed as the profile of the molecular size distribution of the sample. Using the peak positions of narrow molecular weight standards to calibrate the column and instrument under the GPC test condition, and further convert the molecular size distribution of a sample to apparent (also known as calibrant-relative) molecular weight distribution has been widely practiced and known as Peak-Position calibration (*22*), calibrant-relative calibration or conventional calibration. The apparent molecular weights obtained from relative calibration curves can be quite precise when the GPC condition is selected appropriately. To obtain good precisions of mass recovery, MW and MWD is the first step of developing a GPC based method for polymer characterization in industry applications. The mass recovery was determined based on on-line differential refractometer response, the dRI detector calibration constant and the dn/dc determined off-line. The number and weight average PEO-relative molecular weight, the polydispersity index and the mass recovery of the five gelation samples using the best GPC condition identified in this study are listed in Table 3.

The good mass recovery (>98%) and precision on MW averages (<2% standard deviations) of M_n and M_w for all 5 gelatin samples prove the current GPC method is satisfactory. This establishes the GPC results can represent the majority of the sample and provide precise MW and MWD information using the current method. However, when using conventional calibration approach, the accuracy of MW and MWD values can be excellent only when the chemistry and molecular architecture (here the term architecture encompasses both conformational and topological phenomena) of the analytes are identical to the calibration standards. This is not the case in our study, since PEO and gelatins have very different chemistry and might have different molecular structure as well. Therefore, advanced detection systems have to be utilized to obtain accurate or absolute MW and MWD information of gelatins as well as to gain insight of the architecture and thermodynamic status of gelatins in the current solvent system used in the GPC experiments.

Table 3. The PEO-Relative M_n, M_w, PD of Five Gelatin Samples Using Conventional GPC Approach and the Mass Recovery. The Average and Standard Deviation Values Are Based on Three Injections.

GPC-RI	*M_n (kDa)*	*M_w (kDa)*	*Polydispersity (M_w/M_n)*	*Mass recovery (%)*
Gelatin 1	19.6 ± 0.1	123.5 ± 0.1	6.3 ± 0.0	99.8 ± 0.9
Gelatin 2	20.5 ± 0.2	210.4 ± 0.4	10.3 ± 0.1	99.4 ± 0.4
Gelatin 3	9.8 ± 0.0	100.4 ± 0.3	10.3 ± 0.0	98.2 ± 1.1
Gelatin 4	7.0 ± 0.0	34.2 ± 0.1	4.9 ± 0.0	99.0 ± 0.5
Gelatin 5	14.1 ± 0.1	78.2 ± 1.5	5.5 ± 0.1	98.9 ± 0.6

GPC-TD Analysis of Gelatins for MW, MWD, and Viscometry Radius

The aqueous GPC-TD system employed consisted of dual-angle light scattering detector (DALS, low angle 7° and right angle 90°), a differential refractometer (dRI), and a differential viscometer (VISC). Figure 2 represents the typical right angle light scattering (RALS), dRI trace and the molecular weight at each elusion moment obtained using both detectors. The MWD at both high and low molecular regions are very noisy due to the ultra low sample concentration. The calculated values in these two regions can be deviated from the main MWD trendline. Therefore extrapolation was used by the OmniSEC Triple Detection/Light Scattering GPC/SEC software during data processing as demonstrated by the fitted line based on the measured MW as each elution moment to calculate the molecular averages and molecular weight distribution.

The absolute weight average molecular weight (M_W) values of gelatin samples determined by GPC-TD are significantly higher than those determined by the conventional GPC at same elution moments as demonstrated in Figure 3. The M_W versus elution time curve of gelatin 4 was almost a parallel shift from the PEO calibration curve used in the conventional GPC for the main peak region. As expected the obtained absolute weight average molecular weight values of these gelatins determined by GPC-TD as shown in Table 4 were considerably larger than the PEO-relative molecular weights by the conventional GPC (Table 3). This discrepancy can be explained by the fact that the molecular architecture and conformation differences between gelatins and PEO standards when the GPC separation follows the ideal size exclusion separation mechanism. The on-line viscometer along with concentration detection can provide the thermodynamic status of the analyte polymer in terms of intrinsic viscosity [η] to help further investigate the separation mechanism under the selected GPC condition.

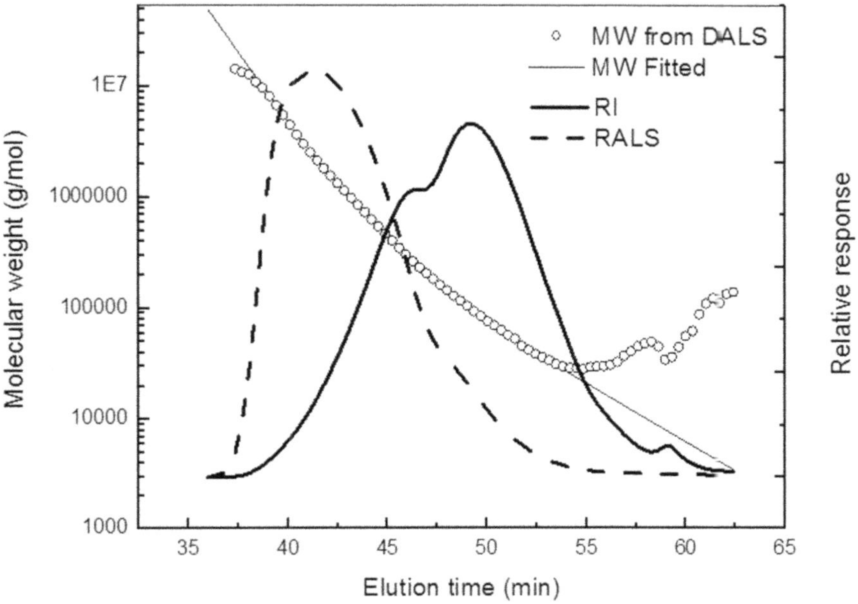

Figure 2. Molecular weight distribution determination of gelatin 2 using GPC-TD overlay with light scattering and dRI traces.

Universal Calibration Deviations

In GPC-TD set-up, the [η] is the ratio of signals from the viscometer and the concentration sensitive detector at each elution moment after correction for interdetector delay and interdetector band broadening. The unit of [η] are volume/mass, dL/g, therefore, the intrinsic viscosity maybe thought of an "inverse density" of the polymer in solution, that is, the more extended (less dense) a polymer is in solution, the higher its intrinsic viscosity will be, and vice versa (*23*). From Einstein's viscosity law, the product [η]M of a polymer chain in solution is directly proportional to the hydrodynamic volume, V_h, of an equivalent sphere. The [η]M versus elution volume plot is well known as universal calibration plot introduced by Grubisic et al (*24*). The universal calibration has proven to be quite robust, polymers of different types merge into a single plot when the GPC condition is fixed in terms of column, solvent, temperature. For example, the universal calibration has been shown to be valid for linear and branched polysaccharides (*25*, *26*) and polyolefins (*27*), and even for denfrimers (*28*) and rodlike polymers (*29*). Therefore, the absolute molecular weight of unknown samples can be calculated using universal calibration without light scattering detector if the universal calibration is valid under the GPC condition being utilized.

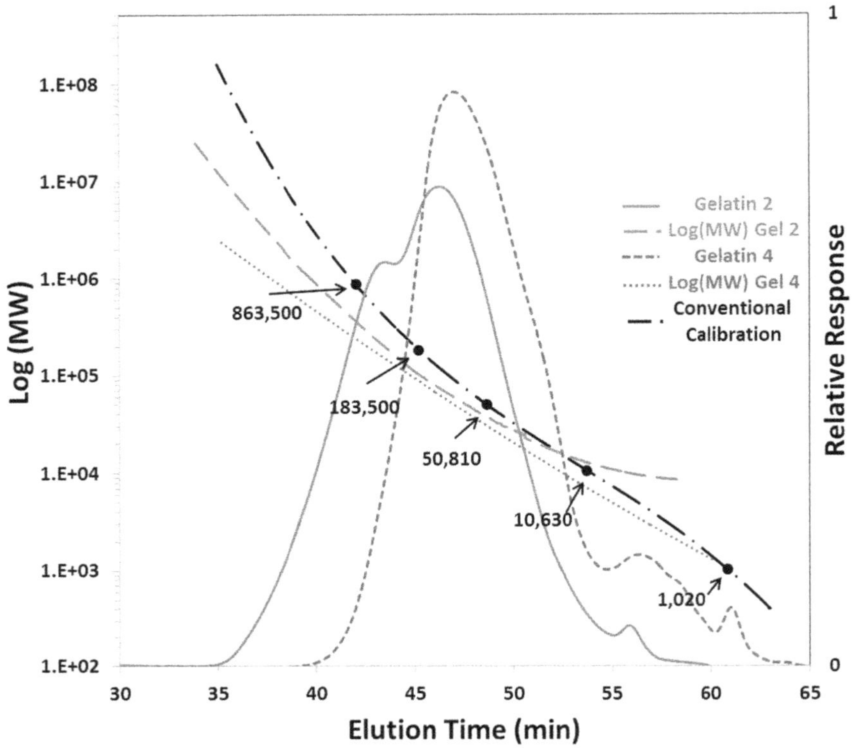

Figure 3. dRI chromatograms and molecular weight –versus-elution time plots of gelatin 2 and 4 from 2-angle light scattering detection with PEO calibration superimposed.

A few exceptions to universal calibration have been discovered over the years if non-ideal size exclusion separation methods were used (*30–32*). If there are other separation mechanisms than the ideal size exclusion involved during the GPC experiments, the molecular weight calculation based on universal calibration can be misleading or erroneous. For example, the [η]M versus elution volume plots, also known as universal calibration plot, of PEO and gelation 2 and 5 (Figure 4) did not completely overlap. The measured polymer volume became large than the value expected based on PEO calibration for the fractions eluted after the peak maximum. Although universal calibration (UC) approach is better than the conventional calibration, the MW cacluated based on UC calibration without using light scattering detector will still be under estimated. The discrepancy between gelatins and PEOs curves suggests the GPC separation employed does not completely follow the ideal size exclusion mechanism. Other non-ideal size exclusion separation mechanism may be present. In other words, under the selected GPC condition, the non-size exclusion effect has not been completely eliminated for these gelatins, and universal calibration could not be used here.

Table 4. Average MW, MWD, Intrinsic Viscosity, and Viscometry Radius of Gelatins Determined by GPC-TD. The Average and Standard Deviation Values Are Based on Three Injections.

GPC-TD	*M_n (kDa)*	*M_w (kDa)*	*Polydispersity (M_w/M_n)*	*[η] (dL/g)*	*R_v (nm)*
Gelatin 1	37.9 ± 0.7	190.5 ± 2.3	5.03 ± 0.15	0.51 ± 0.00	9.96 ± 0.04
Gelatin 2	60.3 ± 0.3	363.3 ± 0.3	6.02 ± 0.02	0.57 ± 0.01	12.43 ± 0.05
Gelatin 3	23.9 ± 1.2	179.7 ± 0.2	7.55 ± 0.38	0.47 ± 0.01	9.50 ± 0.07
Gelatin 4	13.5 ± 0.3	64.3 ± 0.4	4.76 ± 0.11	0.33 ± 0.00	6.32 ± 0.01
Gelatin 5	35.3 ± 0.3	139.3 ± 0.1	3.95 ± 0.03	0.45 ± 0.01	8.98 ± 0.05

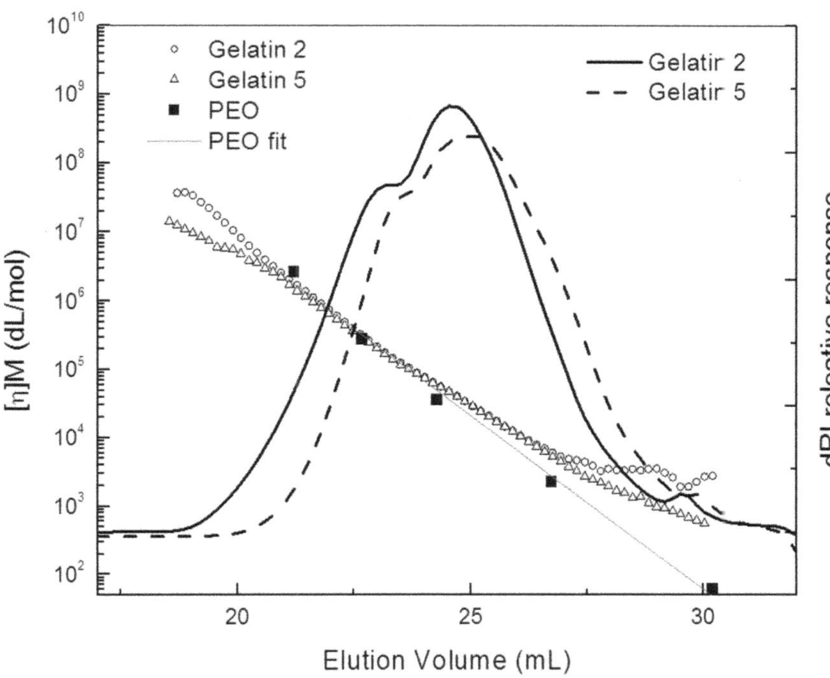

Figure 4. [η]M versus elution volume plots of PEO and gelation 2 and 5 of PEO and gelatin 2 and 5 with RI traces superimposed.

As we are all aware, gelatin is derived from highly crosslinked collagen via acid or base hydrolyses. Inherently, gelatin materials are very complex and have a very broad molecular weight distribution. A mixture of linear and widespread branching in gelatin molecules is expected. Although the NaCl salt, TEA and acetonitrile used in the mobile phase can effectively minimize the ionic interactions and hydrophobic interactions between gelatin molecules and the package materials, it is very difficult to eliminate all the possible interactions during the separation process. These potentially secondary mechanisms other than the ideal size exclusion might explain what we observed in the deviation from the universal calibration.

Therefore, to obtain the absolute MW and MWD information, GPC with on-line light scattering detection is critical and necessary. In this study, the light scattering detector used in the GPC-TD system had only two angles to determine the MW and R_g. Moreover, the low angle detector is very sensity to low level large molecules or particles, the noise level can be potentially high. Therefore, the best solution for this is to use GPC-MALS to determine the absolute molecular weights and R_g of gelatin samples.

GPC-MALS Analysis of Gelatins

Absolute MW from DALS versus MALS

Figure 5 is the comparison of molecular weight vs elution time plots using DALS in GPC-TD system and GPC-MALS. Since the two systems have different dead volume and interdetector delay volume, we shifted the elution volume of the GPC-MALS to match with GPC-TD profiles and used relative response as right Y-axis to better compare the information from two systems. The overlay of both light scattering and dRI traces of the same gelatin sample from both GPC-TD and GPC-MALS is amazingly similar. The molecular distribution determined from both techniques is in good agreement. The comparsion between DALS and MALS using synthetic polymers in organic solvents had been studied, and concluded that the good agreement are expected up to molecular weight 10^6 g/mol and R_g ~60nm (*30*). Here we found in aqueous solution the gelatin sample has a significant portion with MW over 10^6 g/mol, and the absolute molecular weight determined by two techniques still overlapped nicely.

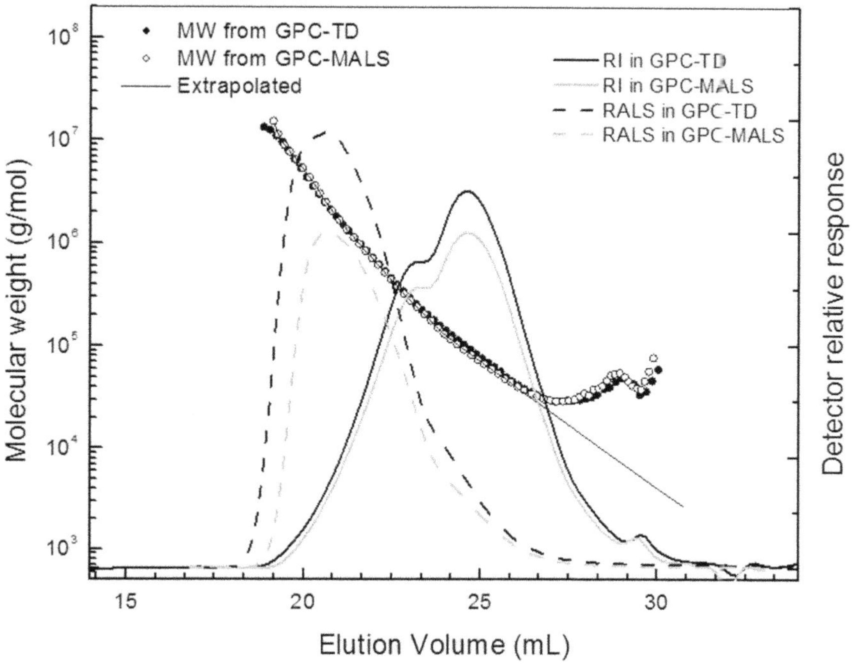

Figure 5. Comparasion of molecular weight vs elution volume plots using DALS in GPC-TD system and GPC-MALS.

R_g from DALS versus MALS

Figure 6 is plotted similarly as Figure 5, but it is the comparasion of R_g vs elution time traces of gelatin 2 using DALS in GPC-TD system and GPC-MALS. Based on light scattering principle, R_g is obtained from the angular dependence of the excess scattering intensity, in other words, it is from the slope of the Debye plot. When the sample contains a mixture of the molecules with different R_g, the obtained R_g value from static light scattering is the z-averaged R_g, therefore, it is very sensitive to the fraction with larger R_g. The agreement of R_g values from two techniques was not as good as that of MW, but both show the elution transition in the main peak region. This is another strong evidence of the anomalous late elution of ultralarge polymer chains under the GPC condition. Since the majority of the sample had R_g below 60nm, this difference is bigger than expected based on the literature study (*30*). Therefore, in order to get reliable thermodynamic status of gelation samples, GPC-MALS is a better approach.

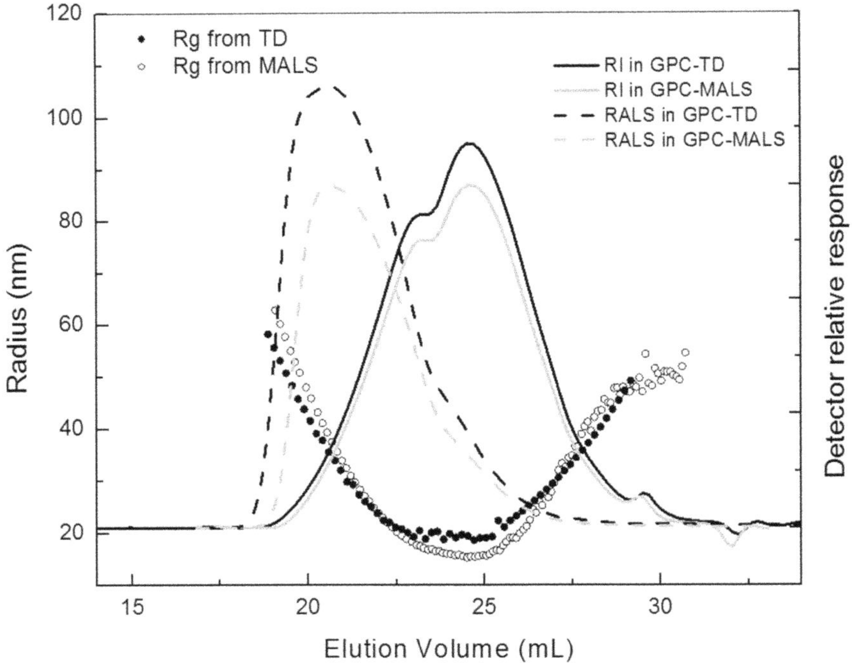

Figure 6. Comparasion of Radius of Gyration vs elution time plots using DALS in GPC-TD system and GPC-MALS. Please note the R_g from TD is based on viscometer measurement and are derived from the Einstein viscosity law for particles in suspension under certain assumptions. See equation 5 and 6.

Data Processing Technique

Due to the exceptionally broad molecular weight distributions of gelatins, it is critical to extrapolate the molar mass trace to include as much information of the sample as possible. Both GPC-TD and GPC-MALS show there is small fraction of high molecular weight components co-eluting with the low molecular weight fraction with elution volume greater than 27 mL (Figure 5). Based on the extrapolated the curve, the MW of the fraction at 30min should be 4200 g/mol, the absolute weight average molecular weight determined by light scattering detectors were about 70 kg/mol. If we assume there were 0.1% large polymers with molecular weight of the 600 kg/mol coeluting with 99.9% 4200 g/mol molecules, the weight average molecular weight of mixture will be 79 kg/mol. Therefore, in order to get better molecular weight determination, the extrapolation is very reasonable.

Without extrapolation the reproducibility of calculated molecular weight and R_g averages had much large standard devation from the same raw data set. The obtained molecular weight averages, polydispersity index and the z-averge R_g from GPC-MALS technique were listed in Table 5. The M_w values for all samples obtained from GPC-MALS and GPC-TD are in good agreement. But for M_n, the difference are bigger than M_w. This difference is mainly due to the data process differences, which was complicated by abnormal size exclusion under the GPC condition. Therefore, due to the exist of the abnormal size exclusion separation, the determination of M_n and PD of gelatins accurately still remains as a challenge.

Table 5. Average Molecular Weight Values and Other Physical Properties of Gelatins Determined by GPC-MALS. The Average and Standard Deviation Values Are Based on Three Injections.

GPC-MALS	*M_n (kDa)*	*M_w (kDa)*	*Polydispersity (M_w/M_n)*	*Conformation plot slope*	*R_{gz} (nm)*
Gelatin 1	49.5 ± 0.5	203.0 ± 2.5	4.10 ± 0.04	0.41 ± 0.01	18.7 ± 1.0
Gelatin 2	66.7 ± 1.1	363.5 ± 3.3	5.45 ± 0.11	0.37 ± 0.00	21.2 ± 0.4
Gelatin 3	38.2 ± 0.3	185.3 ± 1.2	4.86 ± 0.03	0.35 ± 0.00	21.8 ± 1.1
Gelatin 4	13.0 ± 1.0	62.6 ± 1.1	4.86 ± 0.35	0.39 ± 0.05	19.2 ± 1.9
Gelatin 5	45.5 ± 0.6	170.4 ± 0.5	3.74 ± 0.03	0.49 ± 0.01	21.4 ± 1.1

Conclusions

In this study, we have successfully developed a gel permeation chromatographic (GPC) method for separation and characterization of both Type A and Type B gelatin materials. It provides good GPC mass recovery and precise average molecular weight and molecular weight distribution measurement. We performed detailed characterization and comparison of data generated by coupling GPC with triple detection as well as with MALS detection. We also observed some undesirable non-ideal size exclusion chromatographic phenomenon. Nonetheless, both GPC-TD and GPC-MALS have been proved to be effective tools to characterize molecular weights and hydrodynamic and thermodynamic properties, such as intrinsic viscosity, viscometry radius and radius of gyration of gelatin raw materials in the selected mobile phase and temperature. There is good agreement between the two techniques for both the weight-average and the number-average molecular weights. The R_h values calculated by GPC-TD are approximately 2 times smaller than the R_g values determined by GPC-MALS. Although GPC-MALS is better for R_g determination, GPC-TD is valuable for calculating the intrinsic viscosity and viscosity based hydrodynamic radius of gelatin. Conventional GPC calibration were also used to determine the polydispersity and average relative molecular weight values of gelatin molecules.

In summary, this study has demonstrated that GPC-TD, GPC-MALS, and conventional GPC are suitable for characterizing molecular weights, molecular weight distributions, molecular sizes, and solution properties of extremely polydisperse and complex gelatin materials. The knowledge of this detailed molecular information of gelatin raw materials and potential interactions between gelatin and active pharmaceutical ingredients and/or other excipients is tremendously advantageous for processing and quality control. These separation and characterization methods are believed to be able to perform reverse engineering, competitive product analysis, and crosslinking studies. Finally these tools can be used effectively for predicting gelatin-based product performance in the pharmaceutical and nutraceutical industries.

References

1. Schrieber, R.; Gareis, H. *Gelatine Handbook: Theory and Industrial Practice*; Wiley-VCH: Germany, 2007.
2. Wu, C. Simultaneous calibration of size exclusion chromatography and dynamic light scattering for the characterization of gelatin. *Macromolecules* **1993**, *26*, 5423–5426.
3. Scholtan, W.; Lange, H.; Rosenkranz, H.; Moll, F. Bestimmung des Molekulargewichts und der Molekulargewichtsverteilung von Gelatine mit Hilfe der Gelchromatographie und der Ultrazentrifuge. *Colloid Polym. Sci.* **1974**, *252*, 949–970.
4. Meyer, M.; Morgenstern, B. Characterization of Gelatine and Acid Soluble Collagen by Size Exclusion Chromatography Coupled with Multi Angle Light Scattering (SEC-MALS). *Biomacromolecules* **2003**, *4*, 1727–1732.
5. Farrugia, C. A.; Farrugia, I. V.; Groves, M. J. Comparison of the Molecular Weight Distribution of Gelatin Fractions by Size-exclusion Chromatography and Light Scattering. *Pharm. Pharmacol. Commun.* **1998**, *4*, 559–562.
6. Harmon, P. S.; Maziarz, E. P.; Liu, X. M. Detailed characterization of hyaluronan using aqueous size exclusion chromatography with triple detection and multiangle light scattering detection. *J. Biomed. Mater. Res., Part B* **2012**, *100B*, 1955–1960.
7. Liu, X. M.; Gao, W.; Maziarz, E. P.; Salamone, J. C.; Duex, J.; Xia, E. Detailed characterization of cationic hydroxyethylcellulose derivatives using aqueous size-exclusion chromatography with on-line triple detection. *J. Chromatogr. A* **2006**, *1104*, 145–153.
8. Gao, W.; Liu, X. M.; Gross, R. A. Determination of molar mass and solution properties of cationic hydroxyethyl cellulose derivatives by multi-angle laser light scattering with simultaneous refractive index detection. *Polym. Int.* **2009**, *58*, 1115–1119.
9. Zimm, B. H. The Scattering of Light and the Radial Distribution Function of High Polymer Solutions. *J. Chem. Phys.* **1948**, *16*, 1093–1099.
10. Rbii, K.; Violleau, F.; Guedj, S.; Surel, O. Analysis of aged gelatin by AFlFFF-MALS: Identification of high molar mass components and their influence on solubility. *Food Hydrocolloids* **2009**, *23*, 1024–1030.
11. Tromp, R. H.; ten Grotenhuis, E.; Olieman, C. Self-aggregation of gelatine above the gelling temperature analysed by SEC–MALLS. *Food Hydrocolloids* **2002**, *16*, 235–239.

12. American Polymer Standards Corporation. *Light Scattering dn/dc Values.* http://www.ampolymer.com/dn-dc.html (accessed 7/23/18).
13. National Institutes of Health. *Methods for the Biophysical Characterization of Macromolecules: Software, Tutorials, Protocols, and the Workshops from DMAS/LCIMB.* https://sedfitsedphat.nibib.nih.gov/software/default.aspx (accessed 7/23/18).
14. Schuck, P. Size-Distribution Analysis of Macromolecules by Sedimentation Velocity Ultracentrifugation and Lamm Equation Modeling. *Biophys. J.* **2000**, *78*, 1606–1619.
15. Zhao, H.; Brown, P. H.; Schuck, P. On the Distribution of Protein Refractive Index Increments. *Biophys. J.* **2011**, *100*, 2309–2317.
16. Striegel, A.; Yau, W. W.; Kirkland, J. J.; Bly, D. D. *Modern Size-Exclusion Liquid Chromatography: Practice of Gel Permeation and Gel Filtration Chromatography*, 2nd ed.; John Wiley & Sons Inc.: Hoboken, NJ, 2009.
17. Barth, H. G.; Jackson, C.; Boyes, B. E. Size Exclusion Chromatography. *Anal. Chem.* **1994**, *66*, 595–620.
18. Janado, M.; Dubin, P. L. *Aqueous Size-Exclusion Chromatography*; Elsevier Science: New York, 1988; Vol. 40.
19. Mori, S.; Barth, H. G. *Size Exclusion Chromatography*; Springer-Verlag: Berlin Heidelberg, 1999.
20. Normand, V.; Muller, S.; Ravey, J.-C.; Parker, A. Gelation Kinetics of Gelatin: A Master Curve and Network Modeling. *Macromolecules* **2000**, *33*, 1063–1071.
21. Herman, D. P.; Field, L. R.; Abbott, S. The Size-Exclusion Chromatographic Behavior of Synthetic Water-Soluble Polymers on Diol Bonded Phase Supports. *J. Chromatogr. Sci.* **1981**, *19*, 470–476.
22. Striegel, A. M.; Yau, W. W.; Kirkland, J. J.; Bly, D. D. *Modern Size-Exclusion Liquid Chromatography, Practice of Gel Permeation and Gel Fitration Chromatography*, 2nd ed.; John Wiley & Sons, Inc: Hoboken, NJ, 2009.
23. Grubisic, Z.; Rempp, P.; Benoit, H. A universal calibration for gel permeation chromatography. *J. Polym. Sci., Part B: Polym. Lett.* **1967**, *5*, 753–759.
24. Striegel, A. M.; Timpa, J. D. Gel Permeation Chromatography of Polysaccharides Using Universal Calibration. *Int. J. Polym. Anal. Charact.* **1996**, *2*, 213–220.
25. Gaborieau, M.; Castignolles, P. Size-exclusion chromatography (SEC) of branched polymers and polysaccharides. *Anal. Bioanal. Chem.* **2011**, *399*, 1413–1423.
26. Striegel, A. M.; Krejsa, M. R. Complementarity of universal calibration SEC and 13C NMR in determining the branching state of polyethylene. *J. Polym. Sci., Part B: Polym. Phys.* **2000**, *38*, 3120–3135.
27. Striegel, A. M.; Plattner, R. D.; Willett, J. L. Dilute Solution Behavior of Dendrimers and Polysaccharides: SEC, ESI-MS, and Computer Modeling. *Anal. Chem.* **1999**, *71*, 978–986.
28. Temyanko, E.; Russo, P. S.; Ricks, H. Study of Rodlike Homopolypeptides by Gel Permeation Chromatography with Light Scattering Detection: Validity of Universal Calibration and Stiffness Assessment. *Macromolecules* **2001**, *34*, 582–586.

29. Dubin, P. L.; Principi, J. M. Failure of universal calibration for size-exclusion chromatography of rodlike macromolecules vs. random coils and globular proteins. *Macromolecules* **1989**, *22*, 1891–1896.
30. Terao, K.; Mays, J. W. On-line measurement of molecular weight and radius of gyration of polystyrene in a good solvent and in a theta solvent measured with a two-angle light scattering detector. *Eur. Polym. J.* **2004**, *40*, 1623–1627.
31. Bahary, W. S.; Jilani, M. Universal calibration assessment in aqueous gel permeation chromatography. *J. Appl. Polym. Sci.* **1993**, *48*, 1531–1538.
32. Pannell, J. Gel permeation chromatography: the behaviour of polystyrenes with long-chain branching. *Polymer* **1972**, *13*, 277–282.

Chapter 5

Identification of Aqueous Phase Oligomers Formed during Emulsion Polymerization and Their Interfacial Behavior via LC-MS

Tianlan Zhang,* Bryan McCulloch, Wei Gao, and Ralph Even

Dow Chemical Company Northeast Technology Center, Collegeville, Pennsylvania 19426, United States
***E-mail: tianlanzhang@dow.com.**

Interfacial properties of oligomeric molecules play an important role in emulsion polymerization. The radicals generated from water soluble initiators react with monomers in the aqueous phase until the oligomers become sufficiently hydrophobic to partition into the polymer particle phase and carry out the intended further chain propagation. Before reaching the critical composition and molecular weight when oligomers become effectively trapped in the particles, radicals may enter and exit particles readily via diffusion. Better tools and techniques are needed to understand the aqueous phase chemistry and interfacial behavior of aqueous oligomers. In this study, a model emulsion polymerization system was used to generate oligomers with various compositions and chain lengths. Formation of aqueous phase oligomers, as well as the reaction pathways in the early stage of emulsion polymerization, were revealed by LC-MS analysis. Partition coefficients (Log D) of the aqueous oligomers were determined using an LC-MS method to evaluate the amphiphilicity of molecules. It was demonstrated that the composition and amount of aqueous phase oligomers are restricted by their Log D.

Introduction

Free radical emulsion polymerization is widely practiced industrially to produce sub-micron colloidal polymer particles (≈100nm-1µm) dispersed in water, enabling high molecular weight (>100 kg/mol) aqueous-based products with low viscosity and low VOCs for use in a wide range of applications (*1*, *2*). An important part of the overall emulsion polymerization process is its initiation mechanism; a free radical is formed in the aqueous phase where it propagates with dissolved monomer units to form a growing oligomeric radical, the aqueous phase oligoradical. This radical may then undergo any of a variety of types of termination events in the aqueous phase, thus becoming what is referred to as aqueous-phase polymer, or it may grow to a size and composition such that it is energetically favorable to partition out of the aqueous phase and into the particle (*1*). Entry into the particle results in propagation of the radical with monomer which has partitioned into the particle, and this is where the vast majority of polymerization occurs.

While polymerization within the latex particles is similar to bulk polymerization and has been accurately modelled, the aqueous phase kinetics and process of free radical oligomer entry into the particle are very complicated and have been a source of debate, historically (*3–5*). Many efforts have been made to model emulsion polymerization kinetics and polymer properties, but there has been limited success in obtaining detailed information on the nature of the intermediates formed during the polymerization process (*3*, *6*). To date, many of the best efforts have still required rough approximations of the solubility and surface activity for aqueous-phase oligomers. Previous efforts have used mass spectrometry to identify aqueous phase components (*7*); but they failed to identify many of the termination mechanisms present in the aqueous phase and lacked the ability to measure oligomer properties believed to be critical to radical entry into the particle (*8*).

Motivated by this gap in knowledge, we set out to develop the techniques that are required to measure the structure of the oligomers present in the aqueous phase of emulsion polymerization and quantify partitioning behavior of these oligomers. LC-MS using a reverse phase column has been utilized to separate different species and identify oligomer structures. We subsequently attempted to quantify the surface activity of oligomers as a function of their structure and composition by comparing their LC retention times to those of alkyl sulfate surfactants.

Experiment

Preparation of Model Emulsion Polymer Samples

In this study, a low solid, high initiator-to-monomer ratio batch emulsion polymerization reaction was carried out in an attempt to obtain a large amount of aqueous phase terminated oligomers for analysis. Polymers were produced without surfactant to avoid any complications with the characterization of the aqueous phase. Polymerizations were carried out at 85°C using persulfate initiator.

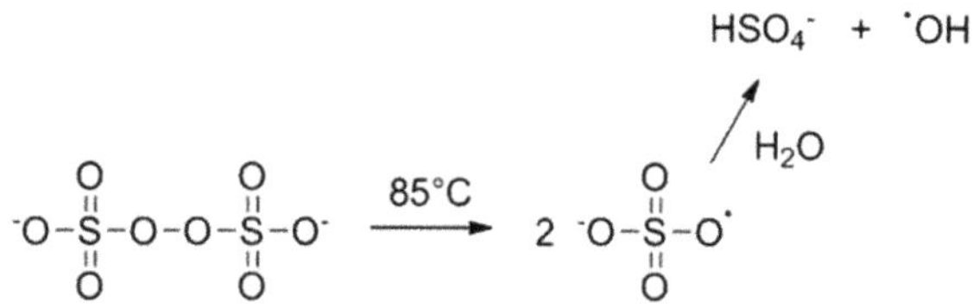

Several acrylic emulsions generated using a model process with methyl methacrylate (MMA) will be discussed in detail. The typical process to produce these materials is as follows. 300g of DI water and 30g of MMA were added to a flask equipped with a water cooled condenser. A nitrogen blanket was applied and the flask was heated with a mantle to 85°C. 4.5g of ammonium persulfate (APS) was added and the reaction was maintained at 85°C. Samples were taken periodically during the reaction and quenched using an aqueous inhibitor solution of 4-hydroxy TEMPO. After 1 hour, the reaction was cooled to room temperature. The reaction had proceeded to full conversion (<5ppm MMA remaining via headspace GC) and the final pH of the latex was 1.7. Other polymers including butyl acrylate (BA), and copolymers with methacrylic acid (MAA) were produced using a similar procedure. To obtain the aqueous phase, free of latex particles, the in-process samples were centrifuged at 543,000 g force for 15 min. The clear aqueous solution was then filtered using a syringe top filter with 0.45um pore size PVDF membrane.

Aqueous Phase Oligomer Analysis by LC-MS

The aqueous phase samples were analyzed using a Waters Acquity binary liquid chromatograph coupled to a Bruker Daltonics micrOTOF-Q mass spectrometer. The LC separation was accomplished by an Agilent Poroshell C18 column with gradient elution of 20mM ammonium acetate at pH 7/acetonitrile at 25°C. High resolution mass spectra were acquired via an electrospray ionization (ESI) in the negative ion mode.

Results and Discussion

Formation of MMA Oligomers in the Aqueous Phase

Oligomers formed in emulsion polymerization may be present in the aqueous phase, adsorbed to the particle surface, or in the latex particles. Therefore, the procedure of isolating different soluble species can be critical. There are a several approaches (*9*) to isolate the aqueous phase oligomers from the latex particles, such as ultrafiltration, ultracentrifugation, serum replacement, or freeze-coagulation. In this study the aqueous phases of in-process samples during the emulsion polymerization were isolated by ultracentrifugation and then directly analyzed by LC-MS without further concentration. The LC-MS chromatograms of in-process MMA aqueous phase samples, at the reaction times of 5 min., 20 min. and 35 min., are shown in Figure 1. Aqueous oligomeric molecules were separated via interactions with a C18 column and a gradient mobile phase. The multiple peaks at different retention times in the chromatograms express the size and composition differences among the oligomers. Clearly, the abundance of aqueous oligomers, as represented by the areas under the LC elution profiles, increases with the reaction time.

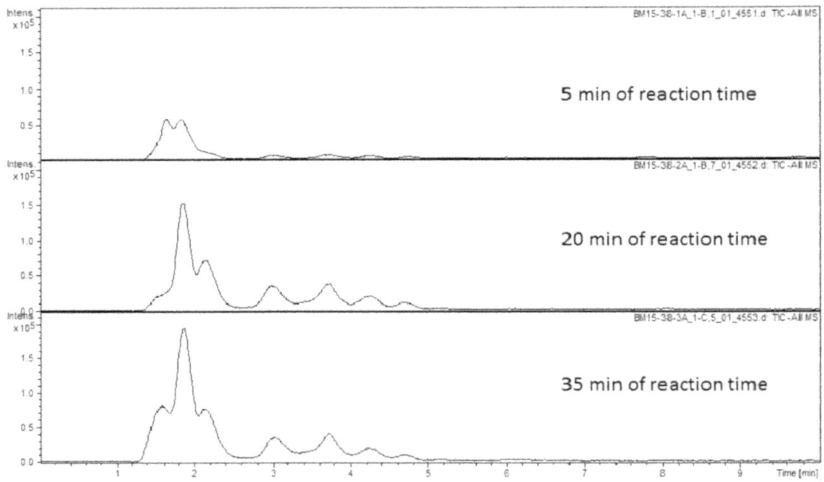

Figure 1. LC-MS Chromatograms of MMA Emulsion at 5, 20, and 35 minutes of reaction time.

To identify oligomer molecules in the aqueous samples, the LC separation was monitored by mass spectrometry. The high resolution mass spectra were acquired in negative ion mode because of the interest in water-soluble polar or anionically charged molecules, which most likely consist of sulfate moieties. The high resolution mass spectrum allowed a mass measurement uncertainty of 5 mDa with the proper mass calibration. To illustrate the interpretation of the mass spectra, the summed spectra from LC retention time of 1.5 to 5.5 minutes are shown in Figure 2. The peaks in the mass spectrum were labeled in Da up to the fourth decimal point.

The empirical formula, or elemental composition, of each of the molecular ion peaks in the mass spectra was determined based on the accurate mass measurement. The molecular structure can be derived from the empirical formula and the reaction chemistry. The measured masses, the empirical formulas of the major components in MMA emulsion, and the measurement uncertainties to the theoretical masses of the empirical formulas are listed in columns 1, 2, 3 in Table 1, respectively. The ion charge state is indicated in parentheses following each empirical formula. The identified molecules determined using the measured masses are listed in column 4. The notation (=) in the table designates unsaturation in the molecule.

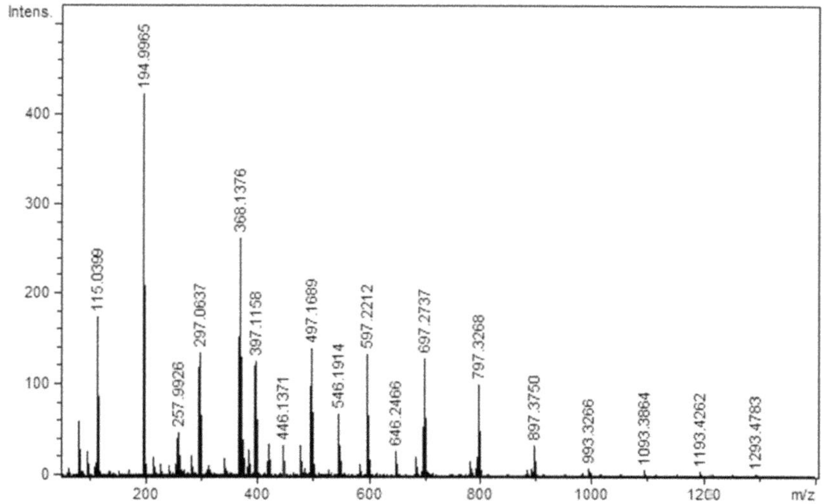

Figure 2. Mass Spectrum of aqueous phase MMA emulsion after 5 minutes of reaction time.

Table 1. Identification of Aqueous Phase Oligomer Ions after 5 Minutes of Reaction Time, Based on Accurate Mass Measurement

Meas. m/z	*Empirical Formula*	*err [mDa]*	*Molecule Identified*
194.9965	$C_5H_7O_6S$ (1-)	-0.4	SO_4-MMA(=)
297.0636	$C_{10}H_{17}O_8S$ (1-)	0.3	SO_4-MMA_2 [a]
366.1219	$C_{14}H_{24}NO_8S$ (1-)	-0.9	SO_4-MMA(=)-4HT *
368.1376	$C_{14}H_{26}NO_8S$ (1-)	-0.9	SO_4-MMA-4HT [b]
395.1013	$C_{15}H_{23}O_{10}S$ (1-)	-2.9	SO_4-MMA_3(=) [c]
497.1689	$C_{20}H_{33}O_{12}S$ (1-)	-0.9	SO_4-MMA_4
546.1914	$C_{45}H_{72}O_{26}S_2$ (2-)	3.2	SO_4-MMA_9-SO_4 (2-) [d]
697.2737	$C_{30}H_{49}O_{16}S$ (1-)	-1.0	SO_4-MMA_6

(note: four structures designated with superscripts *a*, *b*, *c* and *d*)

The molecular structures indicated by superscripts *a*, *b*, *c* and *d* in column 4 of Table 1 are shown in Figure 3. The structures ***a*** and ***c*** are the sulfate-initiated MMA dimer SO_4-MMA_2 and unsaturated trimer SO_4-MMA_3(=), respectively, and result from termination via disproportionation. The structure ***b*** is the SO_4-MMA· radical captured by 4 hydroxy TEMPO right after the sulfate initiation. The structure ***d*** is a doubly-charged nonamer ion, SO_4-MMA_9-SO_4, the recombination product of the two sulfate-initiated MMA oligomer radicals.

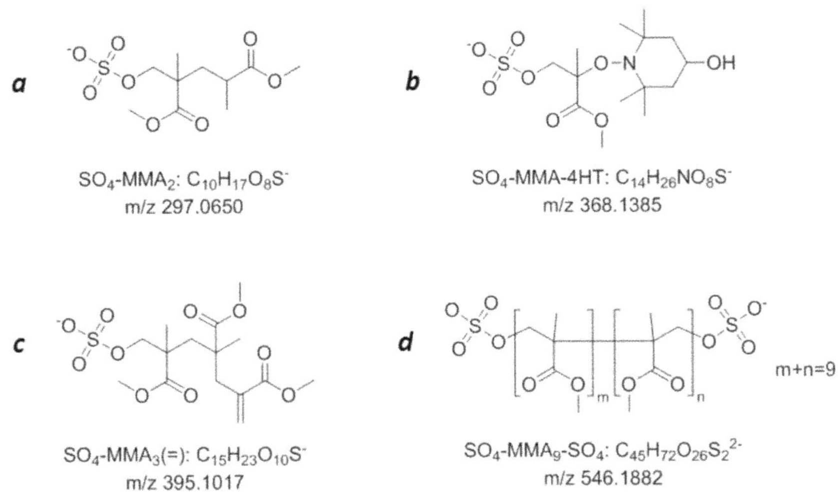

Figure 3. Structure of aqueous species.

The formation of aqueous phase oligomers can be clearly pictured from the aqueous molecules identified by LC-MS analysis. After initiation by radical addition, oligomer radicals begin chain propagation in the aqueous phase. When they reach a critical composition, the oligomer radicals partition to the particle phase of the emulsion. The propagating MMA radicals in the aqueous phase were "accidently" terminated either by radical recombination to form SO_4-pMMA-SO_4 oligomers or by disproportionation, generating both saturated and unsaturated MMA oligomers, SO_4-pMMA. The formation of aqueous phase oligomers initiated by sulfate radical is illustrated in Figure 4.

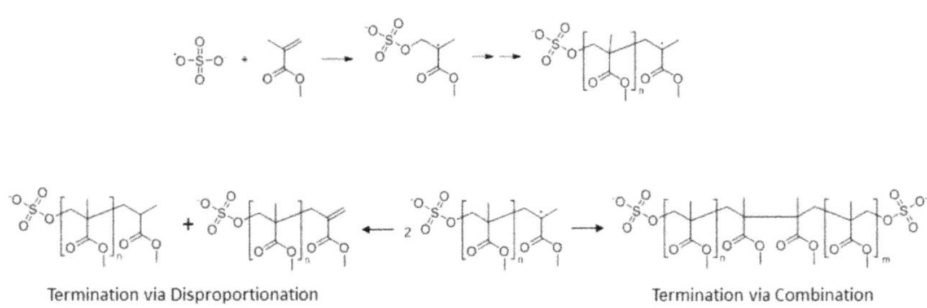

Figure 4. Aqueous phase oligomer initiated by sulfate radical and terminated either through disproportionation or combination.

The extracted ion chromatograms of sulfate-initiated MMA oligomers, sampled at 20 minutes of reaction time, are shown in Figure 5. The oligomers were terminated through disproportionation in the aqueous phase. The top profile is an LC chromatogram of the saturated MMA oligomer SO_4-MMA_m, and the bottom trace is the chromatogram of the unsaturated MMA oligomer, SO_4-MMA_m(=). Each of the peaks in the chromatograms is labeled by the number of MMA repeat units in its corresponding molecule. Molecules up to decamers were found in the aqueous phase for SO_4-MMA_m. LC retention times of the oligomer molecules reflect their relative hydrophobicity. The chromatogram shows that hydrophobicity increases with oligomer chain length growth. Compared to the saturated MMA oligomers, the abundance of unsaturated MMA decreases with reaction time, suggesting that secondary reactions (reinitiation or propagation) may occur.

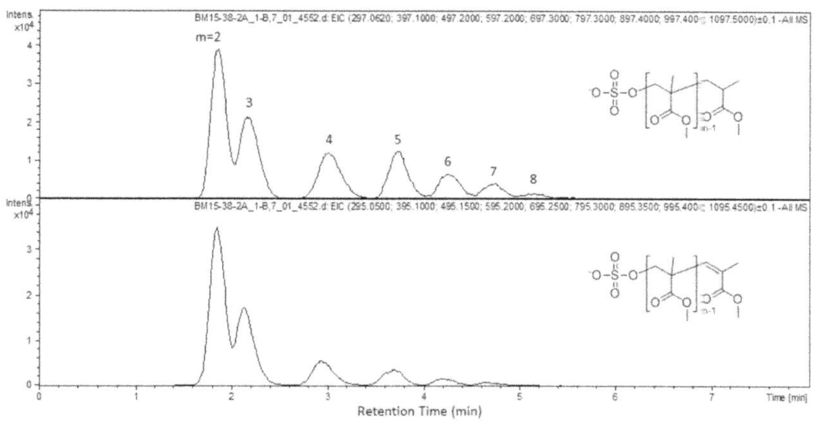

Figure 5. Ion chromatograms of SO_4-MMA vs. SO_4-MMA(=) at 20 minutes of reaction time.

The major oligomer series identified in the analysis of the aqueous phase samples are summarized in Table 2.

Table 2. Aqueous Phase MMA Oligomers Identified by LC-MS after Completion of Reaction

series	*mass range (repeat) Da*	*Oligomer*
1	m/z 297-1197 (100)	SO_4-$(MMA)_n$-H (1-)
2	m/z 295-1195 (100)	SO_4-$(MMA)_n$(=) (1-)
3	m/z 183-1183 (100)	SO_4-$(MMA)_n$/MAA-H (1-)
4	m/z 246/296-946 (50)	SO_4-(MMA)n-SO_4 (2-), n=3-17
5	m/z 393-1493/100	SO_4-$(MMA)_n$-SO_4H (1-), n=2-13
6	m/z 213-913 (100)	SO_4-$(MMA)_n$-OH (1-)

As described above, series 1 and 2 are saturated and unsaturated oligomers of SO_4-pMMA. Since methacrylic acid (MAA) might be a hydrolysis product of MMA, series 3 indicates possible hydrolysis during the emulsion polymerization.

Series 4, 5 and 6 are the recombination products of two propagating MMA radicals, ***I***-pMMA-***I***. Here, ***I*** represents the initiator radical ·OSO3$^-$ or ·OH. Series 4 are doubly-charged oligomers of OSO_3-pMMA-O_3SO^-.

The ratio of disproportionation to recombination termination is estimated to be 2:1 based on oligomer abundances in the aqueous phase by LC-MS and is consistent with the evidence that methacrylates predominately terminate via disproportionation (*10*).

We detected a significant number of small radicals, I-MMA· generated by sulfate initiation of MMA, which had been terminated by the 4HT inhibitor, but only identified a trace amount of I-MMA_2· terminated by 4HT. Based on this, combined with the presence of many much higher molecular weight serum species, we assume that our capture of propagating MMA radicals with 4HT in the aqueous phase was inefficient.

Interfacial Behavior of Aqueous Oligomer

After initiation in the aqueous phase, the propagating radicals become amphiphilic as the chains grow and they start partitioning between the aqueous and polymer phases in the emulsion. The oligomeric radicals may be terminated before entering polymer particles forming amphiphilic oligomers in the aqueous phase. The relative hydrophobicity of these aqueous phase species varies as a function of their composition and size.

LC separation of the aqueous SO_4-MMA oligomers is shown in Figure 6 (top). The chromatogram shows that the hydrophobicity of an oligomer molecule, indicated by its retention time, increases with its chain length. Under the conditions of our study, the longest sulfate-initiated MMA oligomer detected in the aqueous phase consisted of only ten MMA repeat units. It can be concluded that larger oligomers formed in the reaction became so poorly soluble in water that they either precipitated out of the solution due to a lack of water solubility or partitioned into the polymer phase of the emulsion.

The bottom chromatogram in Figure 6 demonstrates the separation of alkyl sulfates in the same LC experimental condition. As expected, the hydrophobicity of alkyl sulfates increases with the number of methylenes in the molecule.

Lauryl sulfate is a commonly used anionic surfactant in emulsion polymerization with a critical micelle concentration (CMC) of ≈8.2 mM at 25°C. Having the same retention time as lauryl sulfate, the sulfate-initiated nonamer, $MMA_9OSO_3^-$, should have similar interfacial behavior, and may displace lauryl sulfate from the surface of the polymer particle. It is also worthwhile to point out that the CMC of hexyl sulfate is about 0.42M at 25°C, having measurable but low surface activity.

Although the formation of aqueous phase polymer is a kinetic-driven chemical reaction, the partition of propagating radicals and oligomers between the aqueous and polymer phases may be predicted by their thermodynamic properties. Amphiphilicity of aqueous phase species can be evaluated using the partition coefficients.

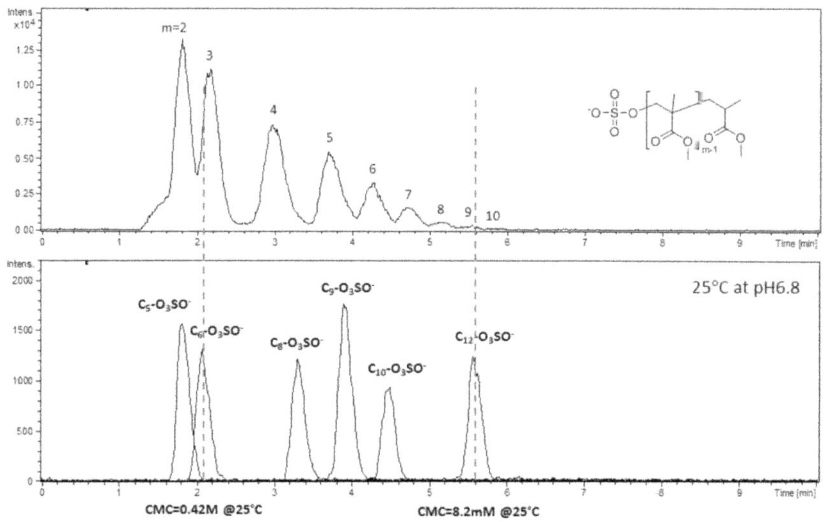

Figure 6. LC-MS chromatograms of MMA oligomers (top) vs. alkyl sulfates $C_{5\text{-}12}\text{-}O_3SO^-$.

Partition coefficient P measures the differential solubility of a molecule in two immiscible phases, n-octanol and water, as commonly applied (*11*),

$$P = \frac{[M]_{oct}}{[M]_{aq}}$$

Here, $[M]_{oct}$ and $[M]_{aq}$ are the concentrations of M in n-octanol and in water, respectively.

If we assume that the monomer-swollen polymer particle in an emulsion system is similar to n-octanol as the hydrophobic phase, amphiphilicity of the oligomer molecule (*5*, *12*, *13*), MMA_mOSO_3H, may be estimated by the calculation of a theoretical partition coefficient between the octanol phase and the aqueous phase. To accommodate both neutral and ionized aqueous oligomer molecules, the distribution coefficient D is used, instead of P

$$D = \frac{[MMA_mOSO_3H]_{oct}}{[MMA_mOSO_3^-]_{aq} + [MMA_mOSO_3H]_{aq}}$$

For the acidic aqueous oligomer MMA_mOSO_3H,

$$Log\ D_{pH} = LogP - Log\left[1 + 10^{(pH - pK_a)}\right]$$

Log D is related to the pKa of the chemical species so it is pH-dependent. Under typical reaction pH conditions in an aqueous environment, alkyl sulfate groups will be anionically charged (pKa < 0).

Liquid chromatography (LC) has previously been used to determine the Log P value of organic molecules for pharmaceutical applications (*14*, *15*). By extension, and while we acknowledge that for our system a strictly defined partition coefficient would be an oversimplification, we propose to use correlation of the LC retention times of aqueous oligomers to those of well-defined surfactants to express the oligomers' hydrophobicity / surface activity as Log D. During the LC analysis, aqueous molecules are separated based on their hydrophobicities. It is accomplished through repartitioning the molecules between the hydrophobic stationary phase of the C18 column and the aqueous mobile phase. The more hydrophobic molecules spend more time in the hydrophobic stationary phase and have longer retention times. The LC retention time is related to Log D of the molecule.

Linear sulfate-initiated oligomers are structurally similar to alkyl sulfates, with a polar head and hydrophobic tail structure, and their amphiphilic behavior is expected to be comparable. By obtaining Log D values for a series of alkyl sulfate molecules, we will be able to correlate their Log D values to the retention times on a C18 column. Using Log D values and the corresponding retention times of alkyl sulfates as a reference, Log D of sulfate-initiated MMA oligomers may be estimated from their retention times under the same LC experimental conditions.

Log D values of alkyl sulfate at pH 5.5 and 7.4 by ACD/Labs prediction software are readily available in the ChemSpider database (*16*). (Log D value at pH 7.4 is often quoted to give an indication of the lipophilicity of a drug at the pH of blood plasma.) The correlation of Log D values to LC retention times from a series of alkyl sulfates is shown in Figure 7.

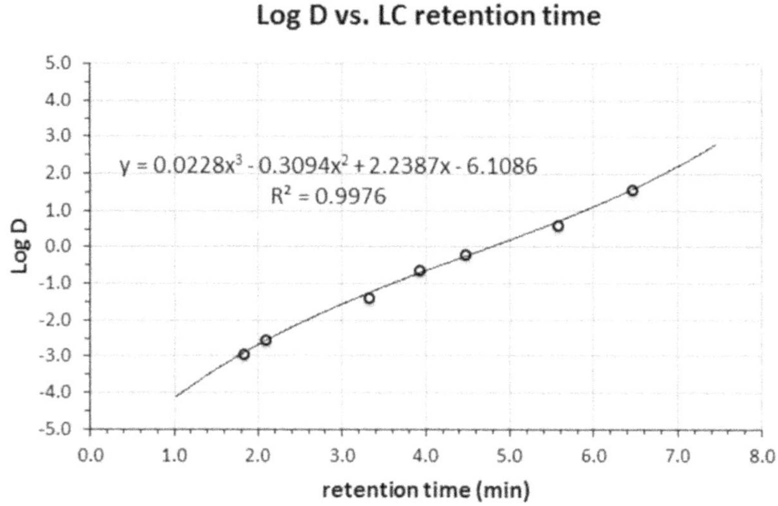

Figure 7. The correlation of Log D values to LC retention times for alkyl sulfates.

By comparing to the plot in Figure 7, Log D of any aqueous molecule can be estimated based on its measured LC retention time. Based on the values of the corresponding alkyl sulfates, Log D values of BA_mOSO_3H, MMA_nOSO_3H and MMA_m/MAA_nOSO_3H, generated in the model emulsion process, are plotted as a function of number of repeat units in Figure 8.

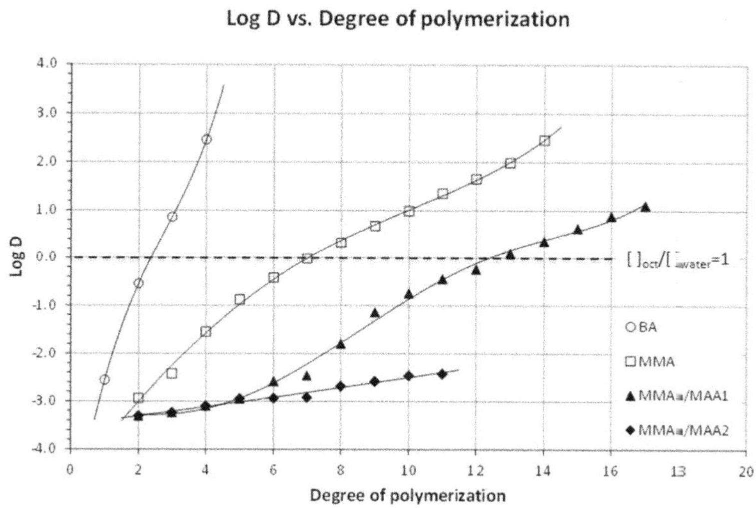

Figure 8. Distribution coefficients (Log D) of various aqueous phase oligomers.

The plot shows that Log D increases with the chain growth, as the oligomer becomes more hydrophobic. Log D of an oligomeric molecule is dictated by the solubility of its monomer composition. Reflective of the difference in water solubility between BA and MMA, Log D of the BA dimer BA_2OSO_3H is higher than the value of the MMA pentamer MMA_5OSO_3H. In the case of the tetramers MMA_3/MAA-OSO_3H, MMA_4-OSO_3H and BA_4OSO_3H, Log D increases significantly from -3.1, -1.6, to 2.5, respectively, because the water solubility of the monomer decreases from MAA, MMA, to BA. Based on these estimations of Log D, MMA_3/MAA-OSO_3H and MMA_4-OSO_3H are soluble in water, while BA_4OSO_3H (with a Log D of 2.5) would be expected to be primarily either precipitated or absorbed onto the polymer particle in an emulsion system.

$MMA_{10}OSO_3H$, having a Log D of around 1, is partitioned 10:1 in the polymer particle phase compared to the aqueous phase. Since decamer has only a 10% possibility of staying in the aqueous phase, oligomers with more than 10 MMA repeat will be mostly absorbed onto the polymer in an emulsion system.

LC chromatograms of two SO_4-MMA oligomers made in slightly different process conditions are shown in Figure 9.

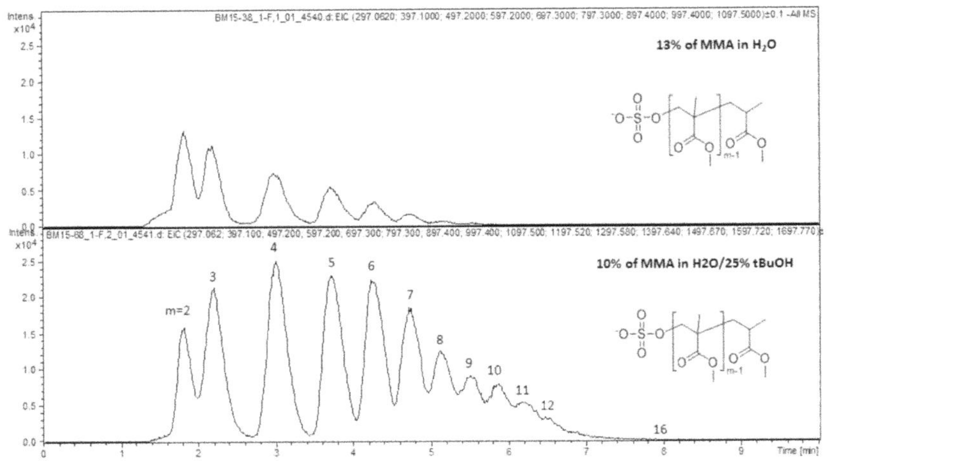

Figure 9. Aqueous phase MMA oligomers formed in 100% aqueous emulsion polymerization (top) vs. those formed in emulsion polymerization in 75% water / 25% butanol (bottom).

The top chromatogram is the separation of SO_4-MMA oligomers, discussed earlier. The oligomers were produced by the emulsion process described in the experimental section. The bottom chromatogram shows the separation of MMA oligomers that were made in the same process but with a continuous phase of 25% tert-butanol in water. Since the continuous phase became much less polar, higher molecular weight MMA oligomer molecules were more soluble. The size and abundance of oligomers increased significantly in 25% tert-butanol emulsion compared to the same reaction products in pure water. This supports the conclusion that the distribution and amount of aqueous phase oligomers are indeed restricted by their Log D.

While LC experiments were conducted at around pH 7, practical acrylic emulsion polymerizations could be carried out over a range of pH. Log D is pH-dependent and, although the above exercise provides very useful insight, the real partition coefficient in emulsion polymerization may have some degree of variation. For the systems where acids of higher pKa are used (ex. MAA), we expect a strong pH dependence in partition coefficient. More LC experiments over a range of pH and temperature will be needed to account for and model actual emulsion polymerization conditions.

Conclusions

The aqueous phase reaction can be monitored by LC-MS during the emulsion process, providing structural information for aqueous phase molecules, as well as aqueous phase oligomer composition. The reaction pathways in the early stage of emulsion polymerization were revealed from the aqueous molecules and radicals identified by LC-MS analysis.

Log Ds of the sulfate initiated oligomers were predicted using the LC-MS method. Partition coefficients can be used to gauge the interfacial behavior of the aqueous phase oligomers and radicals. The composition and amount of aqueous phase oligomers are restricted by their Log D.

References

1. Gilbert, R. G., *Emulsion polymerization: a mechanistic approach*; Academic Press, 1995.
2. Chern, C. S. Emulsion polymerization mechanisms and kinetics. *Progr. Polym. Sci.* **2006**, *31*, 443–486.
3. Maxwell, I. A.; Morrison, B. R.; Napper, D. H.; Gilbert, R. G. Entry of free radicals into latex particles in emulsion polymerization. *Macromolecules* **1991**, *24*, 1629–40.
4. Dong, Y.; Sundberg, D. C. Radical Entry in Emulsion Polymerization: Estimation of the Critical Length of Entry Radicals via a Simple Lattice Model. *Macromolecules* **2002**, *35*, 8185–8190.
5. van Berkel, K. Y.; Russell, G. T.; Gilbert, R. G. Entry in Emulsion Polymerization: Effects of Initiator and Particle Surface Charge. *Macromolecules* **2003**, *36*, 3921–3931.
6. Thickett, S. C.; Gilbert, R. G. Emulsion polymerization: State of the art in kinetics and mechanisms. *Polymer* **2007**, *48*, 6965–6991.
7. Thomson, B.; Wang, Z.; Paine, A.; Lajoie, G.; Rudin, A. A mass spectrometric investigation of the water-soluble oligomers remaining after the emulsion polymerization of methyl methacrylate. *J. Polym. Sci., Part A: Polym. Chem.* **1995**, *33*, 2297–304.
8. Tauer, K.; Deckwer, R. Polymer and groups in persulfate-initiated styrene emulsion polymerization. *Acta Polym.* **1998**, *49*, 411–416.
9. German, A. L.; Van Herk, A. M.; Schoonbrood, H. A. S.; Aerdts, A. M., Latex Polymer Characterization. In *Emulsion Polymerization and Emulsion Polymers*; Lovell, P. A., El-Aasser, M. S., Eds.; John Wiley & Sons Ltd.: Chichester, England, 1997.
10. Nakamura, Y.; Yamago, S. Termination Mechanism in the Radical Polymerization of Methyl Methacrylate and Styrene Determined by the Reaction of Structurally Well-Defined Polymer End Radicals. *Macromolecules* **2015**, *48*, 6450–6456.
11. Vik, E. A.; Bakke, S.; Johnson, D. R.; Verburgh, J. The N-Octanol/Water Partition Coefficient. In *Produced Water 2. Environmental Science Research*; Reed M., Johnsen S., Eds.; Springer: Boston, MA, 1996; Vol. 52.
12. Marestin, C.; Guyot, A.; Claverie Direct Measurement of Oligomers Entry Rate onto Latex Particles in an Emulsion Polymerization. *Macromolecules* **1998**, *31*, 1686–1689.
13. Colombie, D.; Sudol, E. D.; El-Aasser, M. S. Direct Measurement of Oligomers Entry Rate onto Latex Particles in an Emulsion Polymerization. *Macromolecules* **2000**, *33*, 4347–4353.

14. OECD, Partition Coefficient (n-octanol/water), High Performance Liquid Chromatography (HPLC) Method. In *OECD GUIDELINE FOR TESTING OF CHEMICALS*, 1989; Vol. 30.03.89 (Adopted)
15. Paschke, A.; Manz, M.; Schüürmann, G. Application of different RP-HPLC methods for the determination of the octanol/water partition coefficient of selected tetrachlorobenzyltoluenes. *Chemosphere* **2001**, *45*, 721–728.
16. http://www.chemspider.com/ Februrary 27, 2018.

Chapter 6

Characterization of Ultralarge Polymers by Gel Permeation Chromatography: Challenges and Opportunities

David M. Meunier,[*,1] Yongfu Li,[1] and Wei Gao[2]

[1]The Dow Chemical Company, Core R&D Analytical Sciences, Midland, Michigan 48667, United States
[2]The Dow Chemical Company, Core R&D Analytical Sciences, Collegeville, Pennsylvania 19426, United States
[*]E-mail: DMMeunier@Dow.com.

Analysis of ultrahigh molecular weight, or more precisely, ultra large polymers represents one of the most challenging applications for size-based separation by techniques like gel permeation chromatography (GPC). Most of the fundamental obstacles to successful analysis of ultra large polymers can be overcome by proper choice of experimental conditions. Nevertheless, identifying appropriate GPC conditions for linear random flight polymers having a significant weight fraction (e. g., > 3 wt. %) of their molecular weight distribution in sizes exceeding ~300 nm radius of gyration (R_g) remains a challenge. For reference purposes, 300 nm R_g would roughly correspond to the size of a linear polystyrene of ~ 2.4×10^7 g/mol dissolved in tetrahydrofuran at ambient temperatue or a linear polyethylene homopolymer of ~1.0×10^7 g/mol dissolved in trichlorobenzene at elevated temperature.

We have developed a protocol for evaluating the GPC separation of broadly distributed ultra large polymers. We have also developed a strategy for finding ideal, or near ideal GPC conditions for these polymers. We have applied this strategy to a number of ultra large polymer systems, including water soluble polymer materials such as cellulose ethers and polyethylene oxide. One success story, where ideal or near-ideal GPC conditions have been identified, will be shown to illustrate the approach. When ideal, or near-ideal conditions can be identified, GPC coupled with molecular weight sensitive detection provides an accurate molecular weight distribution. A second, more challenging, example will be presented and discussed. This second example will serve to illustrate some of the current limitations toward successful size-based separation and analysis of ultra large polymers. Ideas for overcoming these limitations will also be presented.

Introduction

Gel permeation chromatography (GPC) (*1*), also known as size exclusion chromatography (*2*) is the method of choice for characterizing molecular weight distributions of synthetic polymers. As such, GPC is used heavily in industry for characterization of new materials, enabling structure-property relationships, optimization of polymer processes and solving problems related to polymer molecular weight. Many industrial practitioners of GPC have faced the challenge of separating ultrahigh molar mass polymers. In this chapter, new insights on the key factors influencing the GPC behavior of ultrahigh molar mass polymers are presented and discussed.

Before delving too far into background material on GPC of ultrahigh molar mass polymers, the authors would like to make a key distinction. The term ultrahigh molar mass is somewhat vague considering the mechanism of ideal GPC involves separation by size. For this reason, the authors prefer the term ultralarge as being more descriptive of these challenging to characterize materials. Granted, the term ultralarge is vague without context, but in the authors' experience, defining ideal, or near ideal GPC conditions becomes extremely challenging for linear random flight polymers having one weight percent or more of its mass in sizes exceeding about 300 nm radius of gyration (R_g). To put this into some context, a narrow molecular weight distribution linear polystyrene standard of $2.4x10^7$ g/mol dissolved in THF at ambient temperature would have a radius of gyration of ~300 nm (*3*), while a linear polyethylene chain of about $1.0x10^7$ g/mol dissolved in a good solvent at elevated temperature (e. g., 150 °C) would exhibit a similar R_g. On the other hand, a crosslinked polystyrene nanoparticle having a narrow molecular weight distribution with molecular weight of $2.4x10^7$ g/mol and a volumetric swelling factor of 2 has an R_g of only 50 nm at ambient temperature in THF (*4*), and it is relatively trivial to separate these types of materials by GPC. Thus, ultrahigh molar mass is a somewhat ambiguous term, whereas ultralarge is more precise when the limit of 300 nm is specified.

Like many other topics in chromatography, Giddings performed a thorough analysis of the challenges associated with GPC of polymers of ultrahigh molecular weight (*5*). Key considerations are shear degradation (and more importantly, deformation), non-equilibrium transport between the mobile phase and stationary phase, concentration polarization, stress-induced diffusion, hydrodynamically induced diffusion and the multi-path effect. The aforementioned factors are strongly influenced by experimental variables such as eluent linear velocity, packing particle size, solvent viscosity, sample concentration and packing pore size, to name a few. Virtually all of the key factors identified by Giddings lead to late elution of ultralarge polymer chains. Although it is often difficult to pinpoint the exact mechanism of late elution, it is imperative that one employ a laser light scattering detector to observe any anomalies, as simple inspection of concentration detector chromatograms is not sufficient to diagnose chromatographic non-idealities.

Shear degradation during GPC analysis of various polymers has been reported by many. In virtually all of the previous reports, flow rate was shown to have a significant impact on GPC elution. For example, the apparent size distribution of starch was observed to be highly dependent on flowrate over the studied range of 1.0 to 0.1 ml/minute, however, since no light scattering detector was used, it was difficult to decouple shear degradation from shear deformation (*6*). GPC coupled to multiangle laser light scattering provided unambiguous evidence for shear degradation of the polysaccharide alternan at higher flowrates (*7*). Approaches for minimizing degradation of ultrahigh molar mass polystyrene and polyethylene have also been reported (*8–11*). Additionally, column frits were shown to play a minimal role, or no role at all, in polymer degradation observed during GPC (*12*).

Even after optimization of all experimental variables to reduce or minimize late elution, gaps in availability of appropriate packing materials remain for separating ultralarge polymers by size. Size separation of ultralarge chains requires ultralarge pores. One of the fundamental assumptions in GPC is that liquid within a pore is stagnant, such that there is minimal, or ideally, no flow through a particle. Flow suppression in the pores is governed by the ratio of the pore diameter to the particle diameter, as this ratio controls the flow rate around the particle versus through the particle. As larger pores are needed to separate larger molecules, suppressing flow through the pores becomes a challenge. A polymer chain flowing through a particle via a pore conduit will exhibit delayed elution compared to polymers diffusing in and out of pores, and hence, flowing around the particles.

Many have advocated the use of field flow fractionation techniques for separating ultrahigh molecular weight polymers due to the low shear stresses imparted on polymer chains in the separation channel (*13*). Asymmetrical flow field flow fractionation would appear to have an advantage over GPC for separating highly branched polymers (*14*, *15*) and polymers having only a small weight fraction of ultrahigh molar mass chains (*16*) or microgels (*17*). However, when the entire sample comprises ultrahigh molecular weight linear chains, field flow fractionation techniques become challenging. In field flow fractionation, the injected mass must be kept low for ultrahigh molecular weight polymers to avoid channel overloading and obtain good size separation. Some have succeeded by signal averaging 10 injections of ≤ 1 μg of ultrahigh molar mass polyacrylamide onto the asymmetrical flow field flow fractionation channel (*18*). Although successful, such an approach is hardly practical for routine industrial application.

According to Giddings, "it appears that no insurmountable hurdles exist which would prevent the size-based fractionation of ultrahigh molecular weight polymers by GPC (*5*)." The authors of this chapter agree with this statement, especially for linear random flight polymers of ultrahigh molar mass. In our laboratory, we employ a systematic approach for developing ideal (or near ideal) GPC conditions for separation of ultralarge linear polymers. The details of this approach will be presented and discussed in this book chapter.

Experimental

Materials

All chemicals were purchased from vendors and were used as received. Deionized water (DI Water) was filtered through a 0.2-µm nylon cartridge (EMD Millipore, Billerica, MA, USA) prior to use. Sodium azide, 99.99+%, and bovine serum albumin (BSA) monomer, ~98% were purchased from Sigma-Aldrich. Toluene, HPLC grade was purchased from Fisher Scientific. Polyethylene oxide (PEO) narrow molecular weight standards were obtained from Tosoh. Hydroxypropylmethylcellulose (HPMC) samples were obtained from The Dow Chemical Company. Nominal composition and viscosity of the cellulose ethers are provided in the text. Broad molecular weight distribution PEO materials were obtained from The Dow Chemical Company and are described in Table 1. Bulk weight-average molecular weights (M_W) were determined by flow injection polymer analysis (FIPA) using low angle static light scattering measurements performed at 7° scattering angle and concentration detection by differential refractive index while R_g (the radius of gyration at $M = M_W$) values were estimated from a published conformation plot (*19*). The FIPA experiments are further described below.

Table 1. PEO Materials Used in This Study

	M_W *(g/mol)*	[1]R_g *(nm)*
PEO-1	1.00×10^6	67.7
PEO-2	1.75×10^6	93.8
PEO-3	2.50×10^6	115
PEO-4	3.40×10^6	138
PEO-5	5.70×10^6	187
PEO-6	$\sim 7 \times 10^6$	217

[1] Calculated from conformation plot in reference 19 at $M=M_W$.

Mobile Phase

The GPC eluent was prepared by dissolving NaN_3 in DI water at a concentration of 0.05 wt%, and was filtered by re-circulation through a 0.04-µm nylon cartridge (Pall Corporation, NY, USA) for 4 to 5 hours to remove particulate impurities prior to use. The refractive index of the mobile phase was 1.3365 as measured by a Wyatt Optilab rEX DRI detector at temperature 28°C and wavelength 660 nm.

Sample Preparation

GPC sample solutions were prepared in pre-chilled (4°C) mobile phase to minimize interference from any salt peak. To minimize contributions from chain overlap and concentration polarization, sample concentrations $\leq 0.25C^*$, where C^* is the polymer chain overlap concentration, were required. C^* was estimated as the reciprocal of the intrinsic viscosity of the particular sample. Intrinsic viscosities were estimated from previous literature for HPMC (*20*), or from published Mark-Houwink coefficients for PEO (*21*). Sample solutions were slowly shaken on a flat-bed shaker for 2-3 hours at ambient temperature to dissolve the samples, and then were held overnight in a refrigerator set at 4°C for complete hydration and dissolution. On the second day, sample solutions were shaken again for about 1 hour at ambient temperature and filtered through a 0.45-µm nylon syringe filter (Pall Corporation, NY, USA) prior to injection.

GPC-MALLS-DRI

A Waters 2690 pump equipped with a filter consisting of two layers of 0.2-µm nylon membrane, installed prior to the injection valve, was set at the desired flow rate. The injection was performed by a Waters 2690 autosampler programmed to inject 100 µL of solution. Separation was carried out using either one or two TSK-GEL GMPW columns of either 21.5 or 7.8 mm i.d. x 300 mm length, packed with either 13 or 17-µm particles, having 100Å to 1000Å pores nominal as claimed by the vendor and were operated at 28°C. Specifics on which columns were used are denoted in the text. The detectors were a Wyatt DAWN MALLS detector connected to a Wyatt Optilab rEX differential refractive index (DRI) detector. The Wyatt DAWN was equipped with a red laser operating at a wavelength of 660 nm and operated at ambient temperature, and rEX DRI detector was equipped with a light emitting diode (LED) having the same wavelength as that of DAWN and operated at 28°C.

Calibration

The Wyatt DAWN MALLS was calibrated with HPLC grade toluene filtered through a 0.05-µm syringe filter. The individual light scattering detectors were normalized to the 90° light scattering detector using mono-dispersed bovine serum albumin (BSA) monomer as the isotropic scatter. The inter-detector delay volume was determined by aligning the 90° light scattering and DRI peaks obtained for BSA.

Software and Data Processing

Specific refractive index increments (d*n*/d*c*) of 0.140 and 0.133 were used for the cellulose ethers and PEO, respectively. MALLS data were analyzed using ASTRA 6 software, version 6.1. M_W and R_g values were determined at each SEC elution volume increment. To obtain these metrics, angular variation in scattering intensity from detector angles including 25.8° to 132.2° were least-squares fit to a first-order polynomial according to the Zimm formalism. Custom processing of MALLS data was also performed and further details are discussed in the text.

GPC-Low Angle Laser Light Scattering (LALLS) and FIPA-LALLS

In addition to GPC-MALLS, PEO samples were characterized by GPC-LALLS. GPC conditions were the same as described above. LALLS was performed using a Viscotek Model 302 triple detector array with light scattering detection at 7 and 90 ° scattering angles (although only the 7 ° scattering data were analyzed) and a DRI for concentration detection. Viscotek OmniSEC software was used for data processing. For FIPA experiments, a single 7.8 mm i. d. by 300 mm length TSK-Gel-GMPW column was used to separate polymer from any matrix peaks. The flow rate was 0.2 ml/minute. The LALLS and DRI detectors were calibrated with several PEO standards (Tosoh) ranging in weight average molecular weight from 20,000 g/mol to > 1,000,000 g/mol.

Results and Discussion

Pay Attention to Deborah

In addition to chain scission, ultralarge polymers are susceptible to shear deformation. Elongational stresses act upon polymer chains as they are forced by pressure driven flow to travel through a packed column bed. At some critical strain rate, polymers undergo a coil-stretch transition, leading to anomalous elution in both hydrodynamic chromatography (HDC) (*22*) and GPC (*23*). The Deborah number (D_e) is a useful parameter for assessing GPC conditions. D_e is a dimensionless parameter defined as the ratio of the characteristic time scale of polymer relaxation to the time scale of observation. In the GPC and HDC examples cited above, anomalous elution was observed for experimental conditions where the D_e exceeded 0.5. The expression used for determination of D_e is provided in Equation (1), where τ is the longest relaxation time and $\dot{\epsilon}$ is the elongational strain rate.

$$D_e = \tau\dot{\epsilon} \tag{1}$$

These terms are further defined in equations (2) and (3) below, where C is a constant equal to 1, η is the solvent viscosity, k is the Boltzmann constant, T is the temperature in Kelvins, k_l is a constant of the order 21.68 for randomly packed column beds, Q is the volumetric flow rate, r is the column radius and d_P is the packing particle diameter. It should be noted that the equation for τ applies to linear random flight solutes only.

$$\tau = \frac{C\eta R_g^3}{kT} \quad (2)$$

$$\dot{\epsilon} = k_1 \frac{Q}{d_p \pi r^2} \quad (3)$$

Direct visual evidence of this coil-stretch transition occurring at a Weisenberg number of ~ 0.5 has been reported for fluorescently labeled DNA undergoing extensional flow in semidilute solutions (*24*). The Weisenberg number is defined as the product of the longest relaxation time and the applied strain rate, which is identical to the D_e calculation used by the chromatographers. The only difference being that the DNA experiments were not performed in a packed column bed, whereas the chromatography experiments were performed in a packed bed. However, even the earliest papers involving observation of DNA stretching in extensional flows actually used the term "Deborah Number" to characterize the experimental conditions (*25*). For any rheology purists reading this work, the authors realize that the Weisenberg number and D_e are technically not the same, but a discussion of the differences is beyond the scope of this chapter. The interested reader should consult the literature (*26*). Finally, others have shown that at high D_e, (e. g., De > 1.0), fully stretched DNA can undergo a pinning entanglements as the chains flow through packed beds (*27*). For ultralarge polymers, such events are possible and would certainly lead to anomalous elution in size-based separations such as GPC and HDC.

GPC-MALLS of Cellulose Ethers Containing Ultralarge Fractions

In our previous work on GPC-MALLS of cellulose ether polymers, we presented a systematic approach for defining ideal (or near ideal separation conditions) (*28*). Optimization of sample concentration was discussed in the experimental section. Since the cellulose ether polymers are water soluble polymers, an aqueous eluent was the obvious choice. Additionally, since the cellulose ethers exhibit a lower critical solution temperature (LCST), working at elevated temperature is not a viable option for lowering D_e To evaluate GPC-MALLS conditions for ideal size separation of ultralarge cellulose ethers, molar mass versus elution volume curves were carefully examined as a function of flow rate. Ideal conditions were defined as those giving molar mass versus elution volume curves that were superimposable with that of a broad MWD lower molecular weight sample of the same bulk composition. This idea is illustrated in Figure 1. As can be observed, molar mass versus elution volume curves for the 220,000 cP HPMC (higher molecular weight) sample deviate from that of the 4,000 cP (lower molecular weight) at flow rates of 1.0 and 0.5 mL/minute. A flow rate of 0.2 mL/minute was needed for the 220,000 cP sample to completely superpose M versus elution volume curves.

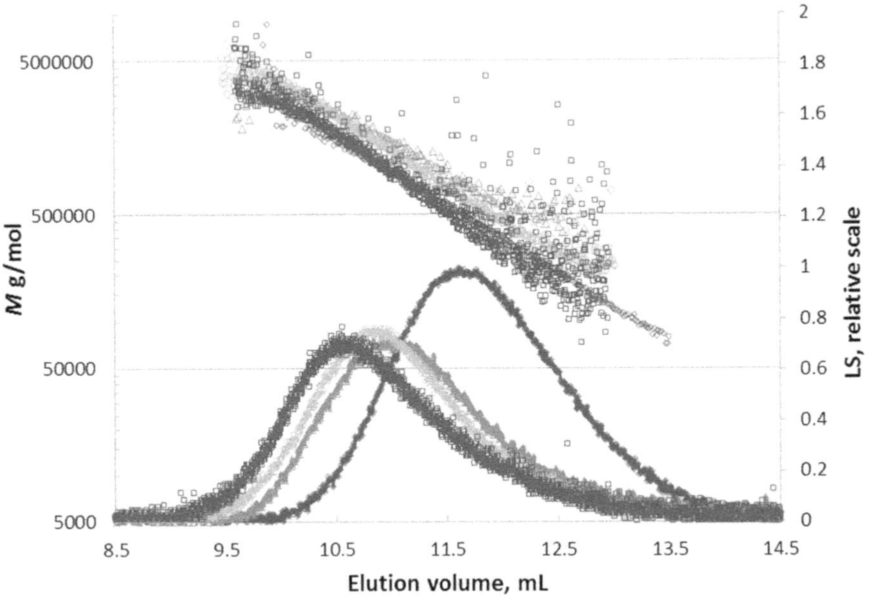

Figure 1. 90 ° light scattering chromatograms and M versus elution volume plots for HPMC having degree of methyl substitution of 1.8 and hydroxypropoxyl molar substitution of 0.13 with 2 wt% solution viscosity of 220,000 cP compared to the same having 2 wt% solution viscosity of 4,000 cP (open diamonds). 220,000 cP data were acquired as a function of flow rate ranging from 1.0 (open triangles) , 0.5 (open circles) and 0.2 mL/minute (open squares) on a pair of GMPW columns each having 7.8 mm i. d. and 300 mm length packed with 17 micrometer particles. The slope of M versus elution volume for 220,000 cP becomes steeper with decreasing flowrate. Reproduced with permission of Elsevier (28).

Given the importance of the coil-stretch transition, further examination of D_e versus R_g at various flow rates is warranted. A plot of D_e versus R_g at different flow rates for the GPC system used for cellulose ether analysis is quite valuable for choosing good conditions. Such a plot is shown in Figure 2, which readily enables one to identify a critical radius of gyration, $R_{g,c}$, as the radius yielding D_e equal to 0.5 at a given flow rate condition. As expected, $R_{g,c}$ increases monotonically with decreasing flowrate, and approaches 295 nm at a flow rate of 0.2 mL/minute, which appears to be the lowest practical flowrate for good light scattering S/N in aqueous mobile phases. An analysis of the influence of $R_{g,c}$ on the goodness of the GPC separation is provided in Table 2. Superimposable M versus elution volume curves were only obtained when the weight fraction of chains exceeding $R_{g,c}$ was essentially zero. Therefore, working at $D_e < 0.5$ for all chains in the sample would appear to be a necessary criterion for good separation. Thus, when developing GPC separations conditions for a sample containing ultralarge chains, experiments to gain knowledge of the largest chains in the sample would be useful to guide proper selection of GPC conditions.

Table 2. Influence of Flowrate on Weight Fraction (W_F) of HPMC Chains Exceeding $R_{g,c}$ and Its Impact on Ideality of GPC Separation Performed on Two 7.8 mm i. d. by 300 mm Length GMPW Columns Containing 17 Micrometer Particles

Flow Rate mL/min	*$R_{g,c}$ (nm)*	*W_F>$R_{g,c}$*	*M vs. EV superimposable with low MW?*
		100,000 cP	
1.0	170	0.05	No
0.5	215	0.01	Yes
		250,000 cP	
1.0	170	0.12	No
0.5	215	0.03	No
0.2	295	~0	Yes
		220,000 cP	
1.0	170	0.10	No
0.5	215	0.03	No
0.2	295	~0	Yes

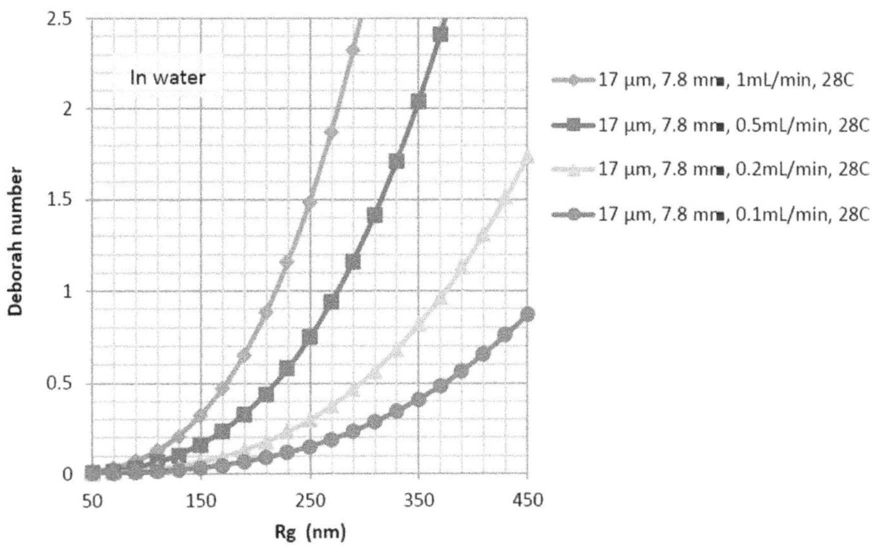

Figure 2. D_e plotted versus R_g at four different volumetric flowrates of water through columns of 7.8 mm i. d. packed with 17 micrometer particles and operated at 28 °C.

Application of Learnings to Ultralarge Polyethylene Oxide

Previous work on characterization of ultralarge PEO by GPC-light scattering identified several key variables in the analysis of these materials (*29*). Flow rate and sample concentration were identified as the most important variables. Analysis of log M versus elution volume plots indicated that ideal GPC conditions remained elusive for the highest molar mass samples analyzed (nominal $M_W \sim 8.0 \times 10^6$ g/mol).

Analysis by GPC-LALLS

GPC-LALLS was applied to the broad MWD PEO materials, and the resulting log M versus elution volume plots are shown in Figure 3. GPC-LALLS analysis was performed at a flow rate of 0.3 mL/minute, which corresponds to an $R_{g,c}$ of ~255 nm. The log M versus elution volume plots are not superimposable. Although the data are not shown, log M versus elution volume of PEO-1 was superimposable with a calibration curve constructed from PEO standards by plotting log M_P values (where M_P is the peak molecular weight provided by the vendor) versus the corresponding peak retention volumes. Thus, PEO-1 appears to be the only sample to elute by an ideal, or near ideal mechanism under these conditions. As PEO M_W increases, the magnitude of deviation increases. This deviation is manifested by later than expected elution of higher molecular weight chains, resulting in a "tipping up" and flattening of log M versus elution volume at longer retention times. This behavior is not due to microgel or branching, since the PEOs studied here comprise linear chains only. The observed behavior would appear to be due to non-idealities associated with characterization of ultralarge chains. Further, LALLS experiments performed in a flow injection type analysis through a GPC column packed with 17 micrometer particles at a flow rate of 0.2 mL/minute yielded bulk M_W values consistent with those determined by GPC-LALLS at 0.3 mL/minute. Therefore, shear degradation does not explain the non-idealities observed in the GPC-LALLS experiments. Given the importance of D_e for separation of ultralarge polymers, it would seem imperative to learn more about the sizes of chains in these ultralarge PEO samples.

Estimation of Largest Sized Chains in PEO by GPC-MALLS

Because the largest sized chains are most susceptible to shear, the highest molecular weight sample, PEO-6 was analyzed by GPC-MALLS at 0.3 mL/minute to gain more insight regarding the size range of the largest chains present in the sample. The results of this analysis are shown in Figure 4. One of the challenges associated with GPC-MALLS of these materials is that extremely low concentrations (~0.1 mg/mL) were used in an effort to minimize non-idealities. As such, S/N can be compromised, especially at the extremes of the chromatograms. One way to improve S/N is to fit the chromatographic peaks. Several commercial programs are available for peak fitting. The built in advanced processing peak fitting capability within GRAMS/AI 8.0 (Thermo electron corporation) was used to fit the light scattering peaks for PEO-6. Prior to peak fitting, any significant spikes were removed using the "Zap" utility within Grams/AI 8.0. Several peak shapes are available in Grams for fitting. For this work, through trial and error, the log-normal function was found to yield excellent fits to the experimental chromatograms as can be observed in Figure 4. Although only one peak fitting example is provided, the fits for the remaining angles were as good, if not better, than that shown. In an effort to learn more about the size range of the largest chains, fitted light scattering chromatograms were used to estimate R_g at the 9.25 mL slice (i. e., where large chains are expected to elute). In the fitting, R_g was allowed to vary as the adjustable parameter in the monodisperse random flight Gaussian coil form factor model (*30*) so as to minimize the sum of squared deviations of the experimentally observed scattering dissymmetry versus the theoretical prediction. The best fit R_g for the 9.25 mL slice was ~ 370 nm. A comparison of experimental versus theoretical prediction shows excellent agreement at the lower scattering angles (Figure 4). Given the prediction of the random flight Gaussian coil form factor of 370 nm, it is clear that the best practical GPC conditions providing the largest $R_{g,c}$ for cellulose ethers (~295 nm) will not be sufficient for PEO-6.

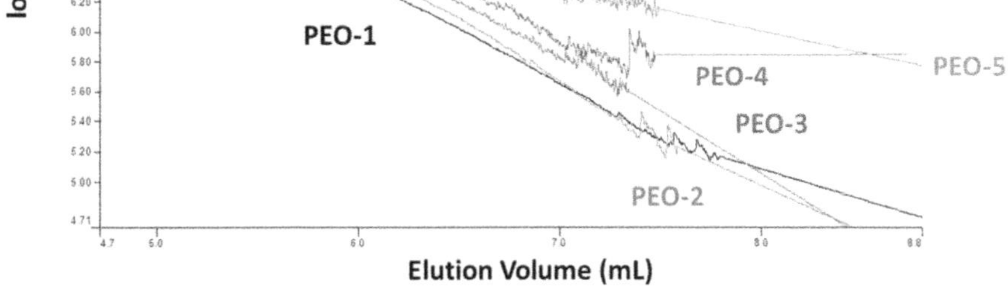

Figure 3. log M_W versus elution volume plots for broad MWD PEO samples of increasing weight-average molecular weight from PEO-1 through PEO-6. GPC flow rate was 0.3 mL/minute on a single GMPW column of 7.8 mm i. d. and 300 mm length packed with 17 micrometer particles. $R_{g,c}$ for this experiment was 255 nm.

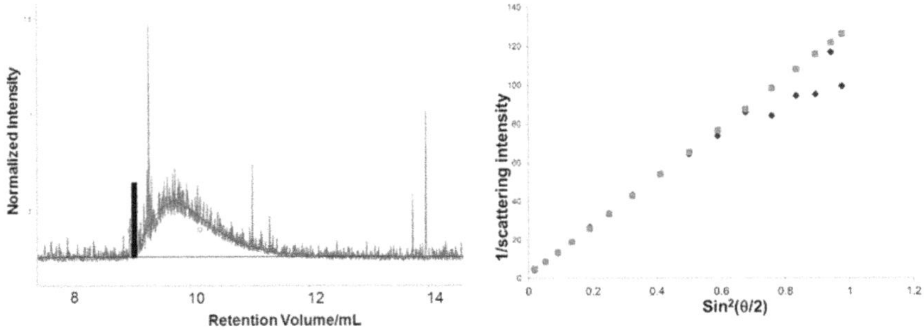

Figure 4. Left panel - light scattering detector chromatogram of PEO-6 for the 26.3° scattering angle (noisy curve) and fitted chromatogram (smooth thin curve). GPC data were acquired at 0.3 mL/minute on a pair of GMPW columns each having 7.8 mm i. d. and 300 mm length packed with 17 micrometer particles. The solid vertical line denotes the slice at 9.25 mL used to estimate the range of large R_g materials in the sample. Right panel – 1/scattering intensity versus $sin^2(\theta/2)$ for each light scattering angle obtained for the 9.25 mL slice compared to the best fit to random coil scattering form factor when allowing R_g to vary as the adjustable parameter in the fit.

Estimation of Largest Sized Chains in PEO Based on the Flory Most Probable MWD

Another approach for estimating the size of the largest chains in PEO-6 was to assume a model for the molecular weight distribution. It is clear that the PEO materials have relatively broad molecular weight distributions. Many processes in the synthesis of polymers produce a Flory most probable molecular weight distribution (*31*). For example, a living polymerization performed in an ideal continuous stirred tube reactor produces a Flory most probable molecular weight distribution due to the residence time distribution of the living chains in the reactor. A Flory most probable molecular weight distribution is characterized by an M_W/M_N of 2. It is not clear whether a Flory most probable molecular weight distribution provides an accurate description of these PEO materials, since good GPC separation conditions have not yet been identified, preventing accurate determination accurate M_W/M_N. Nevertheless, the Flory most probable molecular weight distribution is expected to provide a conservative estimate of the largest sized chains in the system. For PEO-6 an M_N value of 3,500,000 g/mol was used to estimate the Flory most probable molecular weight distribution. The results of this prediction are shown in Figure 5. As can be observed, the weight fraction of polymer becomes vanishingly small at log M of ~ 7.4 corresponding to a molecular weight of ~$2.5x10^7$ g/mol. Based on extrapolation of a previously published conformation plot (*19*), this molecular weight corresponds to an R_g value of ~440 nm.

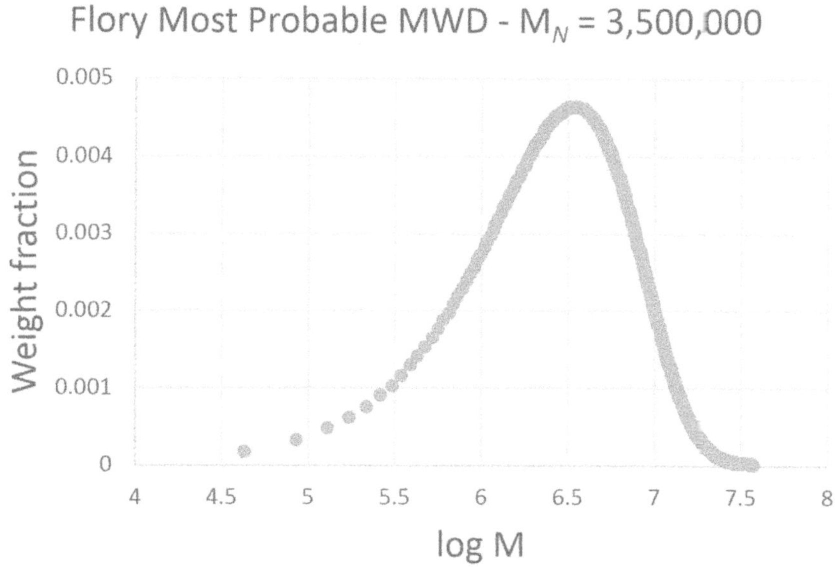

Figure 5. Flory most probable molecular weight distribution prediction for PEO with M_N of $3.5x10^6$ g/mol.

Influence of Column Diameter on D_e

S/N of the MALLS data became unusable at flowrates below 0.2 mL/minute, so further reductions do not appear to be feasible for increasing $R_{g,c}$ beyond the 295 nm found for the cellulose ether separation system. It is clear that the PEO samples contain significant fractions exceeding 295 nm. Further analysis of the factors influencing D_e provides insight regarding further increases in $R_{g,c}$. For example, increasing the column diameter will decrease D_e at a given flowrate leading to an increase in $R_{g,c}$. A comparison of calculated D_e versus R_g is shown in Figure 6 for 7.8 mm versus 21.5 mm inside diameter (i. d.) columns. The main advantage of the larger i. d. column is that it enables operation at lower linear velocity at a given volumetric flow rate. Reduction in linear velocity equates to reduction in elongational strain rate for a given solvent, temperature and packing particle diameter, thus providing access to lower D_e, and hence, larger $R_{g,c}$ all while maintaining adequate volumetric flowrate for good light scattering S/N. For example, at a volumetric flowrate of 0.5 mL/minute, the larger i. d. column provides an $R_{g,c}$ of ~430 nm, which is approaching the largest sizes predicted by the Flory most probable molecular weight distribution. Of course, the tradeoff when using larger i. d. columns is analysis time. At 0.5 mL/minute flowrate through the larger i. d. column, the linear velocity is ~3 times smaller than at 0.2 mL/minute on the smaller i. d. column (the best practical conditions on the smaller i. d. column). Thus, for the same column length, access to increasing $R_{g,c}$ provided by the larger

i. d. column would result in longer run times by about 3X. However, the larger i. d. column contains approximately 7.6 times the amount of packing material than the smaller i. d. column, and hence, about 7.6 times more pore volume for the GPC separation. Thus, it may be possible to use shorter, larger i. d. columns for ultralarge polymer separations.

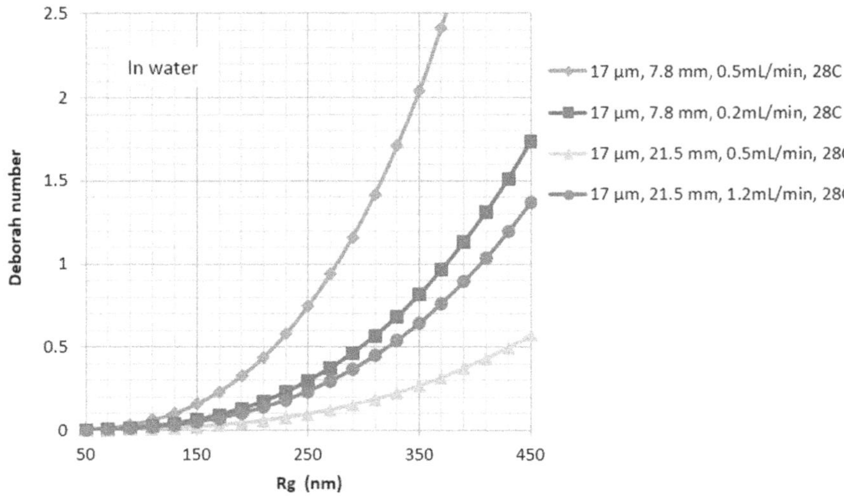

Figure 6. D_e plotted versus R_g for several volumetric flowrates of water through columns of either 7.8 mm or 21.5 mm i. d. packed with 17 micrometer particles and operated at 28 °C.

Application of Larger i. d. Columns for Separation of PEO

Unfortunately, at the time of this work, the 21.5 mm i. d. column was only available in a 13 µm particle size. However, the learnings and comparisons based on work with 13 µm particles are anticipated to be useful for understanding the impact of column radius. To ensure an apples to apples comparison, a smaller i. d. column packed with 13 µm particles was used in the comparison studies. For this work, a volumetric flowrate of 0.5 mL/minute was employed for both columns, and the results are shown in Figures 7 and 8. Results obtained on the smaller i. d. column show significant deviations from ideal GPC separation, and this result is not surprising since $R_{g,c}$ for the experimental conditions employed was only ~195 nm. On the other hand, results obtained on the larger i. d. column show marked improvement versus not only the smaller i. d. 13 µm data, but also the smaller i. d. 17 µm data obtained on two columns at an $R_{g,c}$ value of ~255 nm (Figure 3). Under the flowrate conditions employed (0.5 mL/minute), the larger i. d. column provides an $R_{g,c}$ value of ~395 nm. Under these conditions, PEO1 – PEO4 appear to elute under ideal (or near ideal) GPC conditions, as evidenced by the good superposition of M versus elution volume plots for these materials along with the narrow PEO standards. PEO5 (and presumably PEO6) still appears to exhibit deviations. However, it is clear that increasing $R_{g,c}$ is a favorable move in the direction of ideal separation conditions. For the first time, evidence of total exclusion of a portion of the chains was observed on the large i. d. column, indicating that at least some of the larger solutes were eluting properly. Note, no separate exclusion mode was observed on the smaller i. d. column having the same packing material and operated at the same flowrate. Although it is evident that further increases in $R_{g,c}$ are necessary for successful separation of the PEO materials containing larger sized fractions, it may also be necessary for column vendors to develop new packing materials for size-based separation (as compared to total exclusion) of these chains.

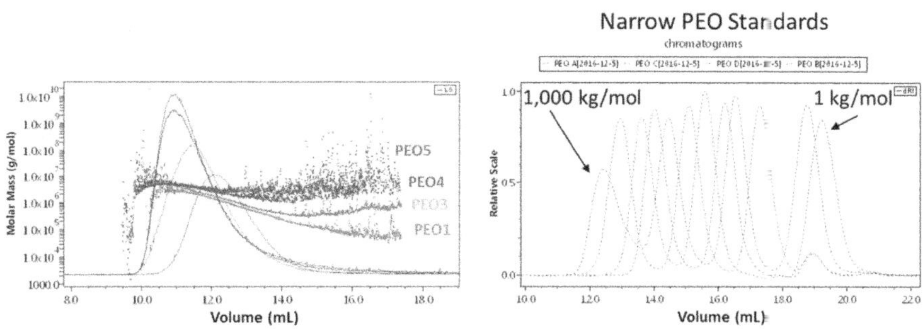

Figure 7. Left panel-GPC chromatograms in order of elution and molar mass versus elution volume plots for PEO5, PEO4, PEO3, and PEO1. Right panel – GPC chromatograms of PEO narrow standards. All chromatograms were obtained using a single 7.8 mm i. d. x 300 mm column packed with 13 μm particles at a flowrate of 0.5 mL/minute to provide an $R_{g,c}$ of 195 nm.

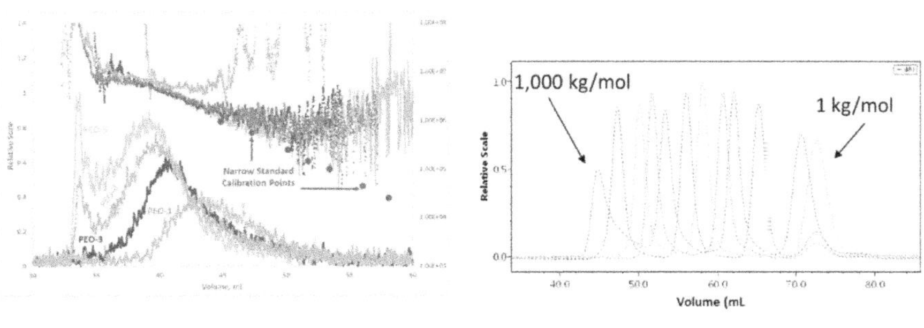

Figure 8. Left panel-GPC chromatograms in order of elution and molar mass versus elution volume plots for PEO5, PEO4, PEO3, and PEO1. M versus elution volume plots for narrow PEO standards are represented by the individual dots. Right panel – GPC chromatograms of PEO narrow standards. All chromatograms were obtained using a single 21.1 mm i. d. x 300 mm column packed with 13 μm columns at a flowrate of 0.5 mL/minute to provide an $R_{g,c}$ of 395 nm.

Toward Further Increases in $R_{g,c}$

Use of lower flowrate and increasing column diameter were shown to have significant impact on the ideality of GPC separation of ultralarge chains. Further analysis of the variables controlling D_e suggest that increases in packing particle diameter, temperature and decreasing solvent viscosity have potential for additional improvements. Although columns packed with 30 µm aqueous compatible particles can be purchased (Shodex), they are only available in a single, relatively large pore size, and as such, are not expected to be useful for broad molecular weight distribution samples. Solvent viscosity represents another potential variable for influencing D_e, and a couple of alternative solvent choices for PEO include chloroform and N, N, dimethylformamide (DMF). DMF has only a slightly lower viscosity compared to water (0.92 versus 1.00 mPA at 20 °C), so would not offer an outstanding advantage. While the viscosity of chloroform is significantly lower than that of water (0.563 versus 1.00 mPA at 20 °C), the safety hazards associated with chloroform need considerable attention before implementation. Increasing the temperature is another mechanism for reducing the solvent viscosity. Temperature increases were not explored as an option to increase $R_{g,c}$ in the previous work on cellulose ethers due to their relatively low LCST in water. Although PEO also exhibits an LCST in water, for the molecular weights in the range of interest here, the LCST was found to be > 90 °C for a 1 wt% solution in dilute aqueous KCl (*32, 33*). Thus, temperature may be a viable knob for increasing $R_{g,c}$ for PEO. Increasing the temperature provides the added benefit of increasing the kT term in the denominator of the expression for τ, resulting in lowering of the longest relaxation time.

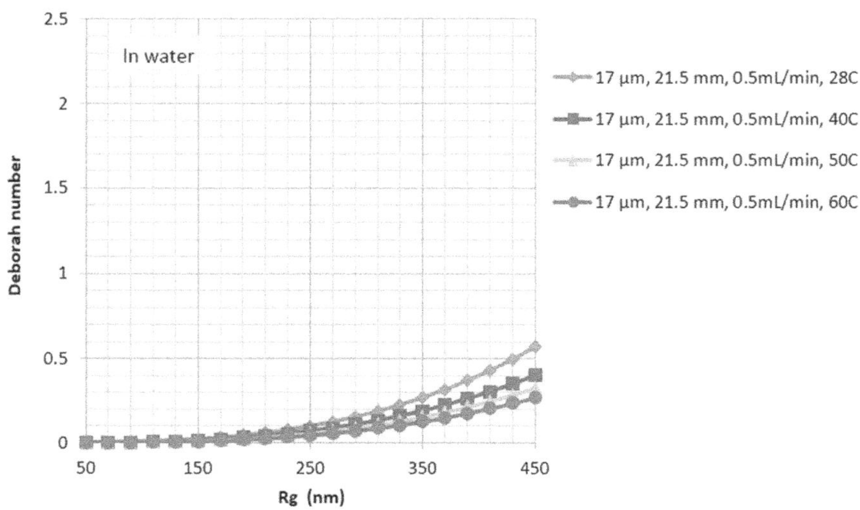

Figure 9. D_e plotted versus R_g for a 0.5 mL/minute volumetric flowrate of water through columns of 21.5 mm i. d. packed with 17 micrometer particles and operated at temperatures ranging from 28 °C to 60 °C.

The impact of temperature on D_e for larger i. d. (21.5 mm), larger particles (17 μm) and a flow rate of 0.5 mL/minute is shown in Figure 9. An increase in temperature from 28 °C to 40 °C provides an increase in $R_{g,c}$ from ~430 nm to ~470 nm. Given the relatively high LCST of PEO in this molecular weight range, it may be possible to work at even higher temperatures to yield $R_{g,c}$ values in the 500 nm plus range.

Column Packing Requirements for Ultralarge Chains

Separation of ultralarge solutes by a size exclusion mechanism requires large pores. Similarly, large particles are required to accommodate large pores. As pores in particles of any size become larger, the fundamental GPC assumption of no flow through the pores becomes suspect. A parameter, ω, is useful to describe the relative flow velocity through the pores versus the interstitial space around the packing particles (*5*), where δ_s is the pore size, d_p is the packing particle size and α has a value of 0.15 for hexagonally closest packed spheres, but is likely closer to 0.2 for randomly packed column beds. Note, an α value of 0.2 was used in the estimations discussed below.

$$\omega = \frac{\delta_s^2}{\alpha^2 d_p^2} \qquad (4)$$

As particle size decreases (i. e. in the direction of more separation efficiency) and pore size increases (as needed for UHMW materials), ω increases. By definition, ω=0 for truly stagnant mobile phase within the pores. In reality, ω must be greater than zero since the pores in a GPC packing particle are not dead ends. For the purposes of defining proper particle sizes and pore sizes, setting a maximum value of ω is somewhat arbitrary. A range of maximum values between 0.1 and 0.3 has been suggested (*5*). Regardless, it is clear that as the flow through the pores increases, the separation mechanism transitions from size-based partitioning to separation by flow (*34*) and eventually to hydrodynamic chromatography (*35*) when ω approaches 1. Nevertheless, equation 4 can be used in conjunction with size estimates of the largest PEO chains to provide guidance on new packing materials for separating these molecules by an ideal GPC mechanism.

Ideal pore diameter, δ_P, for separating solutes of diameter d_s (2 x R_g) is on the order of δ_P/d_s = 2 (*5*). Because flow through the pores is not desired, setting the maximum ω value at 0.1 would appear to be a reasonable choice. Finally, considering our previous definition of ultralarge polymers (R_g > 300 nm), pore sizes and particle sizes required for separating various solute sizes can be estimated. Results of these estimates are provided in Table 3. Using the above assumptions, it would appear that larger particle size, mixed bed packing materials are needed for separating ultralarge polymers according to size. To accommodate 500 nm R_g chains, which is assumed to represent the high molecular weight end of the PEO distribution, ~ 30 μm particles having 20,000 Å pores would be required. Further, such particles should contain a mixture of smaller pores to accommodate separation of the lower molecular weight chains in the broad MWD

PEO samples. Finally, pores larger than 20,000 Å should be avoided to maintain good flow suppression within the pores.

Table 3. Estimated Pore Sizes and Particle Sizes for Separation of Various Sized Solutes. This Analysis Assumes ω = 0.1, α=0.2, and Pore Diameter Is Two Times the Solute Diameter.

R_g *(nm)*	*Pore size (Å)*	*Particle size (µm)*
300	12000	19.0
350	14000	22.1
400	16000	25.3
450	18000	28.5
500	20000	31.6

Conclusions

A systematic approach for defining ideal (or near ideal) size-based separations of linear random flight polymers containing ultralarge fractions has been developed and presented. Evaluation of separations conditions requires a light scattering detector. Comparison of M versus elution volume plots to those of materials known to elute under ideal conditions is an essential component of this approach. Good superposition of M versus elution volume plots occurs when $D_e < 0.5$ for all chains in the system. Under these conditions, accurate molecular weight averages and distributions were obtained for cellulose ether polymers containing ultralarge fractions.

Application of this approach to PEO materials containing even larger sized fractions than the cellulose ether polymers showed there is still need for further development of the technique. Paths toward achieving separations conditions having $R_{g,c}$ near 500 nm were presented. Although experiments have not yet been performed under these conditions, those performed at an $R_{g,c}$ value of ~ 395 nm showed significant progress toward ideal, or near ideal, size-based separation of ultralarge PEO. Under such conditions, an excluded fraction was observed, for the first time, indicating proper elution of the ultralarge chains. Because exclusion was observed, it is clear that larger pores are needed for separation of the ultralarge chains in PEO. Guidelines were offered for making new packing materials to separate the ultralarge chains in PEO.

Besides a lack of suitable packing materials, there appears to be no significant fundamental limit on the applicability of GPC for separating ultralarge linear chains. Application of these ideas to polymers containing branching must be performed with caution. The longest relaxation times for branched polymers are anticipated to be significantly larger than linear chains of the same size. Therefore, it would appear that successful separation of branched chains may require experimental conditions providing significantly lower elongational strain rates. On the other hand, completely different rules may apply to polymers containing branching. Further study on factors affecting ideality of branched polymer elution in GPC is warranted.

References

1. Moore, J. C. Gel permeation chromatography. I. A new method for molecular weight distribution of high polymers. *J. Polym. Sci., Part A: Gen. Pap.* **1964**, *2*, 835.
2. Striegel, A. M.; Yau, W. W.; Kirkland, J. J.; Bly, D. D. *Modern Size-Exclusion Liquid Chromatography: Practice of Gel Permeation and Gel Filtration Chromatography*, 2nd ed.; Wiley: New York, 2009.
3. Van Kreveld, M. E.; Van Den Hoed, N. Mechanism of Gel Permeation Chromatography: Distribution Coefficient. *J. Chromatogr.* **1973**, *83*, 111. R_g and M pairs taken from this paper were plotted and extrapolated to provide estimated M values for $R_g = 300$ nm.
4. Meunier, D. M.; Lyons, J. W.; Kiefer, J. J.; Niu, Q. J.; DeLong, L. M.; Li, Y.; Russo, P. S.; Cueto, R.; Edwin, N. J.; Bouck, K. J.; Silvis, H. C.; Tucker, C. J.; Kalantar, T. H. Determination of Particle Size Distributions, Molecular Weight Distributions, Swelling, Conformation, and Morphology of Dilute Suspensions of Cross-Linked Polymeric Nanoparticles via Size-Exclusion Chromatography/Differential Viscometry. *Macromolecules* **2014**, *47*, 6715.
5. Giddings, J. C. Analysis of fundamental obstacles to the size exclusion chromatography of polymers of ultrahigh molecular weight. *Adv. Chromatogr.* **1982**, *20*, 217–258.
6. Cave, R. A.; Seabrook, S. A.; Gidley, M. J.; Gilbert, R. G. Molecular Structural Differences between Type-2-Diabetic and Healthy Glycogen. *Biomacromolecules* **2009**, *10*, 2245.
7. Striegel, A. M.; Isenberg, S. L.; Cote, G. L. An SEC/MALS study of alternan degradation during size-exclusion chromatographic analysis. *Anal. Bioanal. Chem.* **2009**, *394*, 1887.
8. Aust, N. Application of size-exclusion chromatography to polymers of ultra-high molar mass. *J. Biochem. Biophys. Meth.* **2003**, *56*, 323.
9. Parth, M.; Aust, N.; Lederer, K. Molecular Characterization of Ultrahigh Molar Mass and Soluble Fractions of Partially Cross-Linked Polyethylenes. *Int. J. Polym. Anal. Char.* **2003**, *8*, 175.
10. Aust, N.; Parth, M.; Lederer, K. SEC of Ultra-high Molar Mass Polymers: Optimization of Experimental Conditions to Avoid Molecular Degradation in the Case of Narrow Polystyrene Standards. *Int. J. Polym. Anal. Char.* **2001**, *6*, 245.
11. Striegel, A. M. Observations Regarding On-Column, Flow-Induced Degradation During SEC Analysis. *J. Liq. Chromatogr. Relat. Technol.* **2008**, *31*, 3105.
12. Striegel, A. M. Do column frits contribute to the on-column,flow-induced degradation of macromolecules? *J. Chromatogr. A* **2014**, *1359*, 147.

13. Malik, M. I.; Pasch, H. Field-flow fractionation: New and exciting perspectives in polymer analysis. *Progr. Polym. Sci.* **2016**, *63*, 42.
14. Otte, T.; Pasch, H.; Macko, T.; Bruell, R.; Stadler, F. J.; Kaschta, J.; Becker, F.; Buback, M. Study of the abnormal late co-elution phenomenon of low density polyethylene in size exclusion chromatography using high temperature size exclusion chromatography and high temperature asymmetrical flow field-flow fractionation. *J. Chromatogr. A* **2011**, *1218*, 4257.
15. Podzimek, S.; Machotova, J.; Snuparek, J.; Vecera, M.; Prokupek, L. Characterization of molecular structure of acrylic copolymers prepared via emulsion polymerization using A4F-MALS technique. *J. Appl. Polym. Sci.* **2014**, *131*, 40995.
16. Mes, E. P. C.; De Jonge, H.; Klein, T.; Welz, R. R.; Gillespie, D. T. Characterization of high molecular weight polyethylenes using high temperature asymmetrical flow field-flow fractionation with on-line infrared, light scattering, and viscometry detection. *J. Chromatogr. A.* **2007**, *1154*, 319.
17. Dubascoux, S.; Thepchalerm, C.; Dubreucq, E.; Wisunthorn, S.; Baysse, L.; Kiatkamjornwong, S.; Nakason, C.; Bonfils, F. Comparative study of the mesostructure of natural and synthetic polyisoprene by size exclusion chromatography-multi-angle light scattering and asymmetrical flow field flow fractionation-multi-angle light scattering. *J. Chromatogr. A* **2012**, *1224*, 27.
18. Leeman, M.; Islam, M. T.; Haseltine, W. G. Asymmetrical flow field-flow fractionation coupled with multi-angle light scattering and refractive index detections for characterization of ultra-high molar mass poly(acrylamide) flocculants. *J. Chromatogr. A* **2007**, *1172*, 194.
19. Devanand, K.; Selser, J. C. Asymptotic behavior and long-range interactions in aqueous solutions of poly(ethylene oxide). *Macromolecules* **1991**, *24*, 5943.
20. Keary, C. M. Characterization of METHOCEL Cellulose Ethers by Aqueous SEC with Multiple Detectors. *Carbohydr. Polym.* **2001**, *45*, 293–303.
21. Brandrup, J.; Immergut, E. H.; Grulke, E. A. *Polymer Handbook*, 4th ed.; John Wiley & Sons Ltd.: Hoboken, NJ, 1999; p VII-32.
22. Liu, Y.; Radke, W.; Pasch, H. Coil-Stretch Transition of High Molar Mass Polymers in Packed-Column Hydrodynamic Chromatography. *Macromolecules* **2005**, *38*, 7476.
23. Uliyanchenkoa, E.; Schoenmakers, P. J.; van der Wala, S. Fast and efficient size-based separations of polymers using ultra-high-pressure liquid chromatography. *J. Chromatogr. A* **2011**, *1211*, 1509.
24. Hsiao, K-W.; Sasmal, C.; Prakash, J. R.; Schroeder, C. M. Direct observation of DNA dynamics in semidilute solutions in extensional flow. *J. Rheo.* **2017**, *61*, 151.
25. Schroeder, C. M.; Babcock, H. P.; Shaqfeh, E. S. G.; Chu, S. Observation of Polymer Conformation Hysteresis in Extensional Flow. *Science* **2003**, *301*, 1515.

26. Poole, R. J. The Deborah and Weissenberg numbers. *Rheo. Bull.* **2012**, *53*, 32.
27. DeLong, C. D.; Hoagland, D. A. Imaging of Individual Polymers Undergoing Flow in a Bed of Small Spheres. *Macromolecules* **2008**, *41*, 4387.
28. Li, Y.; Shen, H.; Lyons, J. W.; Sammler, R. L.; Brackhagen, M ; Meunier, D. M. Size-Exclusion Chromatography of Ultrahigh Molecular Weight Methylcellulose Ethers and Hydroxypropyl Methylcellulose Ethers for Reliable Molecular Weight Distribution Characterization. *Carbohydr. Polym.* **2016**, *138*, 290. https://doi.org/10.1016/j.carbpol.2015.11.003.
29. Porsch, B.; Welinder, A.; Körner, A.; Wittgren, B. Distribution analysis of ultra-high molecular mass poly(ethylene oxide) containing silica particles by size-exclusion chromatography with dual light-scattering and refractometric detection. *J. Chromatogr. A* **2005**, *1068*, 249.
30. Debye, P. Molecular-weight Determination by Light Scattering. *J. Phys. Colloid Chem.* **1947**, *51*, 18–32.
31. Flory, P. J. *Principles of Polymer Chemistry*; Cornell University Press: Ithaca, NY, 1953; pp 318–320.
32. Khoultchaev, K. K. H.; Kerekes, R. J.; Englezos, P. The Role of Poly(ethylene oxide)-Water Solution Phase Behaviour in the Retention of Fibre Fines and Clay. *Can. J. Chem. Eng.* **1997**, *75*, 161.
33. Florin, E.; Kjellander, R.; Eriksson, J. C. Salt effects on the cloud point of the poly(ethylene oxide)+ water system. *J. Chem. Soc. Faraday Trans.* **1984**, *80*, 2889.
34. Guttman, C. M.; DiMarzio, E. A. Separation by Flow. *Macromolecules* **1970**, *3*, 681–691.
35. Small, H. Hydrodynamic Chromatography A Technique for Size Analysis of Colloidal Particles. *J. Colloid Interface Sci.* **1974**, *48*, 147–160.

Chapter 7

Development of Asymmetrical Flow Field Fractionation with On-line Advanced Detections for Particle Size Distribution Analysis of Silica Colloidal Particles

Wei Gao,[*,1] Jamie Cohen,[2] Francis Acholla,[3] and Wenyu Su[4]

[1]Analytical Sciences, Corporate R&D, The Dow Chemical Company, Collegeville, Pennsylvania 19426, United States
[2]Chemical Science, Corporate R&D, The Dow Chemical Company, Midland, Michigan 48674, United States
[3]Dow Electronic Materials, Newark, Delaware 19713, United States
[4]Statistic group, Corporate R&D, The Dow Chemical Company, Midland, Michigan 48674, United States
[*]E-mail: weigao@dow.com.

Consistent particle size distributions (PSD) of silica particles are believed to be critical in providing consistent removal rates and defect results when the silica particles are formulated into slurries and used in chemical mechanical planarization (CMP) processes. This study focuses on development and evaluation of a new approach for the characterization of particle size distributions of colloidal silica particles. A special emphasis is given to the average particle size and small mode quantification. The goal is to identify small lot-to-lot variations between well-controlled silica abrasive particle samples that show little variability in typical PSD variables when measured using commonly practiced particle size characterization techniques.

Asymmetrical Flow Field Flow Fractionation with online multi-angle light scattering, dynamic light scattering and differential refractive index detectors (AFFFF-MALS-DLS-dRI) can measure the particle size, concentration and size distribution of a Klebosol® (*1*) 1501-50 slurry within a single one hour run using the developed method. The detailed method, data process approach and the meaning of about thirty indices obtained by this method are explained in this study. The acceptance of these indices for measurement of lot-to-lot variation was assessed based on the measurement of 15 Klebosol® 1501-50 slurry lots with less than 2% variations based on the values of the certificate of analysis provided by the raw material supplier. It was concluded that AFFFF-MALS-dRI is capable of differentiating subtle lot-to-lot differences in terms of concentration, particle molar mass,size averages for the whole sample and the level of the small mode, etc. AFFFF-MALS-dRI is expected to become a powerful tool for the characterization of abrasive particles in the CMP industry for better quality control as well as for fundamental studies.

1. Introduction

1.1. CMP and the Role of Metal Oxide Particle Size in CMP Polishing Slurries

Chemical mechanical planarization (CMP) is widely used in various industrial applications ranging from optical lens polishing to storage media and semiconductor manufacturing. In the semiconductor industry, CMP is a widely used technique for the planarization of metal and dielectric films to accomplish multilevel metallization (MLM). Rapid advances in the microelectronics industry require a decrease in the size of microelectronic devices, and the fabrication of these small feature devices without defects requires significant improvements in the CMP process.

An important characteristic of any CMP process is the overall removal rate (RR) of the surface material being polished. Removal rate is typically described in terms of Preston's law (*2, 3*):

$$\frac{dh}{dt} = k_P \sigma_0 V_r$$

where dh/dt is the average thickness removal rate, σ_0 is the applied pressure, V_r is the average relative velocity of the polishing particles relative to the work piece, and k_P is the Preston coefficient, which is a function of several parameters including reaction kinetics, solids concentration, and surface properties. The Preston equation is an empirical relationship that postulates the proportionality of the removal rate to the applied pressure σ_0 and the relative velocity V_r of the surfaces in contact. It has been demonstrated that, in oxide CMP, the chemical content of the polishing slurry is responsible for softening of the material comprising the film or layer, whereas mechanical forces help achieve the required planarization and uniformity through removal of this softened layer at some desired rate. The abrasive action in CMP is mostly provided by the sub-micrometer size abrasive particles as they flow in between the pad and the wafer surface under applied pressure.

The magnitude of the mechanical action is in turn determined by the concentration and the size of the abrasive particles, among other things. During polishing, complex mechanical and chemical interactions occur between the wafer (e.g. silica wafers), polishing particles (ceria and/or colloidal silica particles), aqueous media (CMP slurry solution formulated with chemicals to enhance chemical erosion), and pad. Optimal removal rate and smooth surface finish are achieved by the synergistic effect of the chemical and mechanical forces encountered in the CMP process.

There are many studies conducted and reported in the open literature to propose, test and model the CMP mechanism and process. The particle size and particle size distribution have been identified as critical parameters to the CMP process. For example, it was observed that certain contamination of larger particles not only created surface damage but also changed the material removal rate (*4*). In a recent paper, Suratwala et al (*5*) reviewed previous related studies and used global material removal rates that were experimentally measured and calculated using the Ensemble Hertzian Gap model to test hypotheses related to the material removal mechanisms. The results of their experiments and simulations provide a number of insights regarding optical polishing: (1) the size dependence of the tail end of the PSD of the slurry strongly influences the roughness of the polished surface; (2) only a small contact area between the work-piece and pad occurs at any given time; (3) material removal with ceria particles likely occurs both by chemical and plastic mechanisms (*6*); (4) the transition from chemical to plastic removal occurs at a critical load per particle ($P>P_{crit}$ $5x10^{-5}N$), and (5) the single particle removal depth does not depend on particle size, however, the particle contact length increases with particle size. Some of these concepts can be applied to semiconductor polishing and are certainly dependent on substrate, material layer, and pad among other input parameter variables

The removal rate of a particular type of slurry is governed by a number of factors associated with the particle and the slurry, but it has been observed that the particle size and the particle size distribution can have a strong effect. In some cases, the particle size can be proportional to the removal rate and in others it can have an inverse relationship, depending on the pH and material being polished. It can also be a source of variation that can cause the removal rate to be unpredictable if there are modes within the overall particle distribution that vary from batch to batch. Since the removal rate is measured in ångströms per minute, strict control over the slurry properties is required in order to obtain the precision needed to planarize at surface roughness values on the order of ångströms. Moreover, when the particle size distribution is bimodal, the particle size distribution effect on removal rate can be even more complicated. In real industrial application, there are also many other parameters that affect removal rate, and even the reported relationship between particle size and RR might not agree. For example, Chandrasekaran et al (*7*) observed the removal rate correlates primarily with the slurry mean particle-size distribution (D50). Lu et al (*8*) used monodispersed particles to study particle size and morphology effects on removal rate. They found that the removal rate of copper wafers increased with silica size when slurries contained the same number of particles, but it decreased with size when a constant weight of the solid abrasives was used. In recalculating these data, it was established that the polishing rate was independent of the silica size for slurries with the same total surface area of dispersed abrasives.

In industrial formulation, CMP slurries are usually formulated so that the abrasive particles have the same percent weight instead of the same total surface area. Therefore, the raw material qualification in terms of the PSD of the abrasive particles is one of the essential steps in ensuring the consistent removal rate of the final formulated CMP slurry. Klebosol® (*1*) 1501-50 colloidal silica abrasive particles are made with a water-glass process and maintain excellent stability in their liquid medium at the pH used during synthesis, however, they are not monodispersed colloidal silica particles (Figure 1). They have an average size of 50 nm as calculated from the specific surface area derived from titration data.

The objective of this study is to evaluate the measurement capability of Asymmetrical Flow Field Flow Fractionation (AFFFF) technology for colloidal silica particles characterization using Klebosol® 1501-50 as model metal oxide abrasive particles.

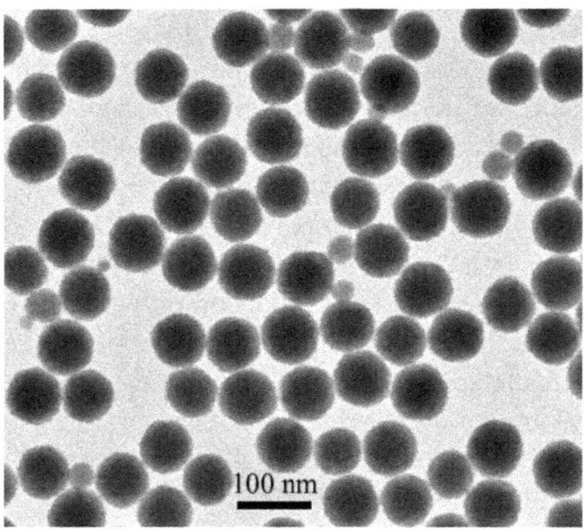

Figure 1. TEM image of Klebosol® 1501-50 Particles.

1.2. Field Flow Fractionation Related Techniques and Their Applications for Silica Particle Size Characterization

Traditionally, both instrumental and chemical methods are used for characterizing the size of colloidal silica. Chemical procedures include gas adsorption, titration, preferential precipitation, and the rate of particle dissolution. Instrumental methods include electron microscopy, x-ray crystallography, laser diffraction, laser light scattering, turbidity-absorbance, sedimentation and nuclear magnetic resonance spectroscopy, etc. Methods that can directly observe the particle size distribution of colloid silica are microscopy methods, such as transmission electron microscopy (TEM), scanning electron microscopy, and atomic force microscopy. However, these methods are not practical for most researchers and are not ideal analytical tools in the quality control of colloidal silica for several reasons. To obtain useful micrographs, the particles must be thoroughly dispersed on a microscope grid. Only an extremely small portion of a sample will be observed, so several replicate measurement must be made to obtain a meaningful statistical distribution of particles. The particles must be measured and counted according to size, which is a tedious procedure when done manually and not trivial when an image analysis program is used. Transmission electron microscopy is best used as a supplemental tool to other methods for occasional check of the samples (*9*). In order to measure the particle size distribution, different separation techniques with advanced detections have been explored, for example, size-exclusion chromatography and hydrodynamic chromatography with multidetector have been investigated for the characterization of string-of-pearls colloidal silica (*10*).

Field Flow Fractionation (FFF) is recognized to be the most versatile technique for submicron particle separation due to the large choice of physical fields. The FFF separation range covers one nanometer to tens of micrometers. However, one FFF condition might not cover this full range and the separation range of a single experiment depends on the particular technique, separation mode and detailed method used (*11*, *12*). Most often different sub-techniques of FFF are presented according to the nature of the physical force used. Thus flow- (Fl-FFF or 4F), sedimentation- (Sd-FFF), thermal- (Th-FFF), magnetic- (Mg-FFF), split-flow thin cell- (SPLITT-FFF), acoustic-(Ac-FFF), electrical- (El-FFF), di-electrophoresis-FFF (DEP-FFF also named dielectrophoresis-FFF) or pressure- (P-FFF) can be quoted (*13*). Among these sub-FFF techniques, the most used are Fl-FFF, Sd-FFF, Th-FFF, El-FFF and SPLITT-FFF. They are also the only techniques that have commercial instruments available.

Asymmetrical Flow Field Flow Fractionation (AFFFF, or A4F, AF4) is a type of Fl-FFF, which has many similarities to chromatography except it does not use any packing material and there are no phase equilibria. The open channel gives rise to relatively low pressures, low shear and gentle sample treatment during fractionation. The fractionation mechanism of AFFFF was originally described by Wahlund (*14*) as illustrated in Figure 2. In brief, the bottom of the AFFFF-channel (accumulation wall) consists of an ultrafiltration membrane with a given molecular weight cut-off which allows some of the carrier liquid entering the channel inlet to exit through the accumulation wall (the crossflow). The resulting 'field' gives rise to a convective transport of sample components toward the accumulation wall. The field induced transport is, in turn, counteracted by the Brownian motion of sample components. At steady-state, each size fraction in the sample has established a characteristic concentration distribution from the accumulation wall. The characteristic distance from the accumulation wall for a size fraction, at a given cross-flow rate, depends on its diffusion coefficient. Size fractions with higher diffusion coefficients (smaller hydrodynamic size) are on average situated further away from the accumulation wall than size fractions with lower diffusion coefficients (larger hydrodynamic size). As the axial flow in the channel is laminar, the flow velocity is highest in the center of the channel and decreases toward the accumulation wall. Size fractions distributed further away from the accumulation wall experience a higher axial flow velocity and are, thus, transported more rapidly out of the channel than size fractions which are in closer proximity to the accumulation wall. Hence, AFFFF fractionates samples based on differences in diffusion coefficient (or conversely hydrodynamic size), and sample diffusion coefficients can be directly determined from elution times when using constant cross-flow (*15*).

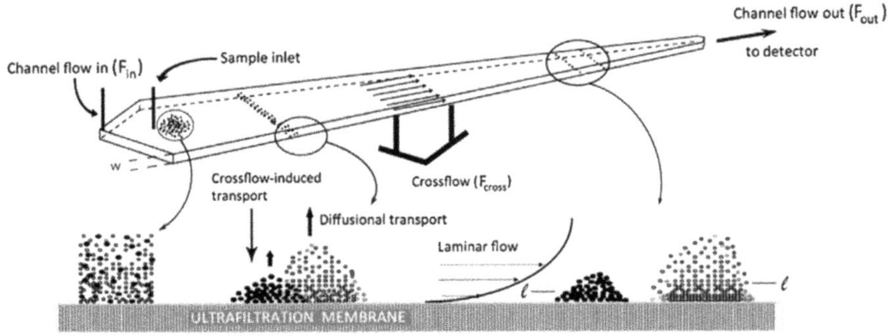

Figure 2. Principle of AFFFF. Fractionation of a mixture of two particles with different size(s). A homogeneous mixture is injected at sample inlet, relaxed and focused. At start of elution the two particle populations start to migrate with different velocities. At the end of the channel two zones have become resolved. Filled (darker color) symbols indicate large particles while open (grey color) symbols indicate small particles. w is channel thickness and l is the center of gravity distances of particle populations from the ultrafiltration accumulation wall. (Reprinted with permission from Ref. (16)).

2. Experiments

2.1. Sample Information of the Klebosol™ 1501-50 Lots

The Silica colloidal samples in this study were in two groups. One group was 15 lots of Klebosol® 1501-50, and the other group was 15 formulated slurries from the CMP slurry group within Dow Electronic Materials. The Certificate of Analysis (CoA) for each Klebosol® 1501-50 lot listed the Specific surface area (SSA), mean particle size (MPS), pH, wt% Silica content and Specific gravity (Sg) (Table 1).

2.2. Materials and Sample Preparation

Particle size standards were purchased from Thermo Scientific. Ammonium acetate and 25% Ammonium Hydroxide solution were purchased from Fisher Scientific. The mobile phase was 20mM Ammonium acetate buffer at pH 10, which was prepared using purified water (MilliQ water purification system with resistivity below 18.2 mΩ·cm at 25 °C), the pH was adjusted to 10 using ammonium hydroxide solution.

The particle size standards and samples were all diluted 100-fold using mobile phase prior to AFFFF analysis. The diluted samples were directly injected into the AFFFF system without further filtration.

Table 1. CoA Values of 15 Klebosol® 1501-50 Lots

Lot	*Ssa (M^2/G)*	*Mps (Nm)*	*Ph*	*Wt% Silica*	*Sg*
1	51.000	83.8	10.806	30.06	1.20532
2	49.300	83.8	10.839	30.04	1.20456
3	N/A	85.2	N/A	N/A	N/A
4	N/A	85.4	N/A	N/A	N/A
5	50.109	84.4	10.795	30.037	1.20507
6	51.255	82.6	10.76	29.9388	1.20492
7	50.209	81.9	10.768	29.9519	1.20468
8	51.645	83.6	10.763	30.0194	1.20525
9	49.778	83.5	10.84	30.0807	1.2049
10	51.087	81.4	10.767	30.0933	1.20523
11	50.214	83.9	10.823	30.182	1.20507
12	51.894	84.2	10.762	30.0981	1.2049
13	50.037	81.6	10.762	30.1271	1.20506
14	50.908	80.7	10.774	30.0648	1.20538
15	52.000	83.3	10.782	29.9739	1.20437

2.3. Instruments

The AFFFF flow regulation was controlled by Eclipse 3+ (Wyatt Technology, Santa Barbra, CA). The separation channel dimensions were 15.2 cm in length and from 2.15 to 0.3 cm in width with a 350 μm thickness spacer. For all experiments, the membranes used were regenerated cellulose with a 10 kDa MW cutoff (Wyatt Technology). Flows were delivered with an Agilent Technologies 1200 series isocratic pump equipped with a micro-vacuum degasser. All injections were performed with an autosampler (Agilent Technologies 1200 series). Multiple on-line detectors were connected in series to characterize the fractions eluting from the AFFFF channel. The detection train consisted of a variable wavelength ultraviolet/visible spectrophotometer (UV) (SP8480 XR Scanning detector, Spectra-Physics, USA), a multi-angle laser light scattering (MALS) detector (DAWN HELOS II from Wyatt Technology) with dynamic light scattering measured at 142.6° (at detector position number 16) (Mobius from Wyatt Technology), and a differential refractometer (RI) detector (Optilab rEX, from Waytt Technology). Data were collected and processed by Astra 6.1.2.76 software (Wyatt Technology).

2.4. Multi-Angle Light Scattering (MALS) Calibration and Normalization

The 90-degree MALS detector was calibrated using HPLC grade toluene (Fisher Scientific). The detectors at other angles were normalized using the 100 nm particle size standard from Thermo Scientific.

2.5. Asymmetrical Field Flow Fractionation (AFFFF) Method for Klebosol® 1501-50 Fractionation and Data Acquisition

The mobile phase used for AFFFF analysis was 20 mM ammonium acetate solution at pH 10. The optimized AFFFF method used for sample analysis was established with detector flow at 1 mL/min, injection flow at 0.2 mL/min, and the detailed time table as listed in Table 2. 50 μL of freshly prepared slurry samples were injected into AFFFF system for characterization. The MALS data collection rate was one second per data point. The DLS data interval was set to 5 sec. Mobile phase was also injected into the AFFFF system to obtain blank injection signals for RI blank baseline subtraction.

Table 2. Time Table of the Optimized AFFFF Method

Start Time (min)	*End Time (min)*	*Duration (min)*	*Mode*	*Cross flow start (mL/min)*	*Cross flow end (mL/min)*
0	1	1	Elution	0	0
1	2	1	Focus	-	-
2	5	3	Focus+inject	-	-
5	6	1	Focus	-	-
6	29	23	Elution	0.5	0.05
29	34	5	Elution	0	0
34	35	1	Focus+inject	-	-
35	40	5	Elution+inject	0	0

3. Results and Discussion

3.1. Sample Handing and Its Effect on Analysis

For particle size analysis, sampling is a critical step to ensure the portion being analyzed can represent the whole sample. Therefore, during the sampling process, shaking the samples well before preparation and injection to AFFFF system was found to be critical. Since the AFFFF system contains an auto-sampler to make precise injections for the fractionation, a large number of samples can be prepared as a group and the autosampler tray can be used to run the analysis automatically.

However, during the long run, the particles might settle in the sample vial and cause measurement issues. Both the main population (Figure 3) and the small mode (the insert of Figure 3) can be affected by settlement over time. The main particle peak can be lowered by more than 10% after sitting for 7 hours.

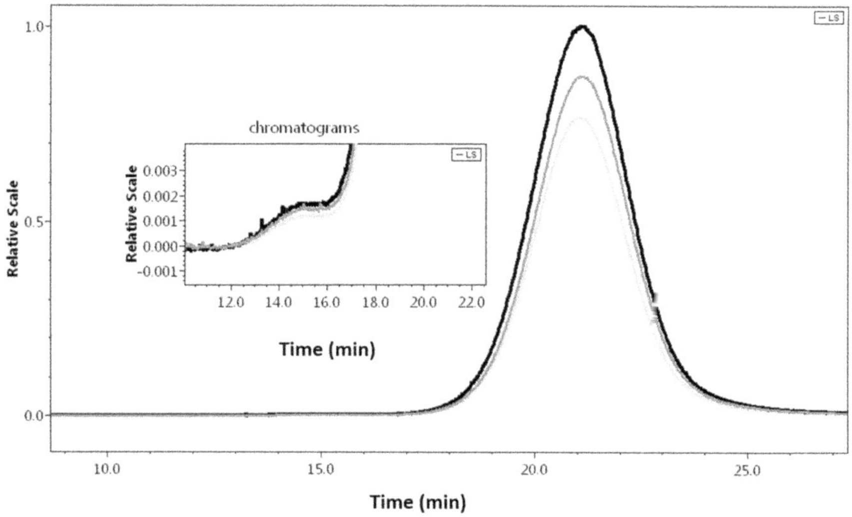

Figure 3. AFFFF Fractograms of a typical sample after seating in the autosampler for different time. Black traces- 0 hours, gray traces: 7 hours, light gray traces: 14 hours. The insert is the zoomed-in view of the small mode. The injection volume was 10 μL.

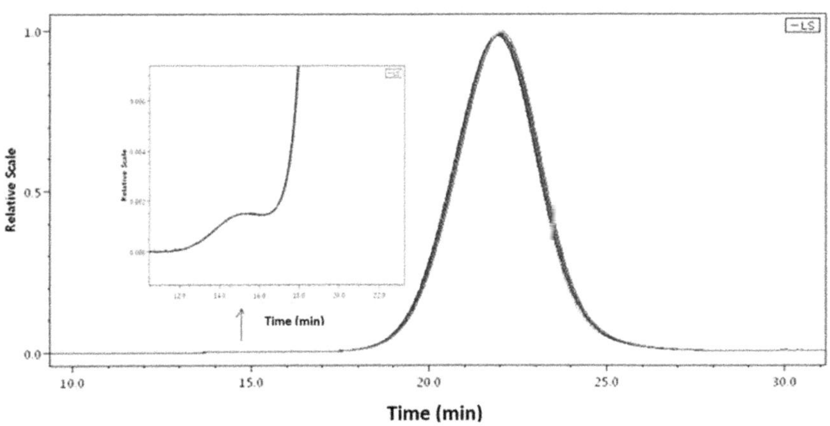

Figure 4. AFFFF Fractograms of ten continuous injections from the lot 1 of Klebosol® 1501-50 colloidal silica abrasive particles. The injection volumes were 10 μL.

*The sample was shaken by hand before the first injection, and the following injections were shaken by the autosampler 40 mins before the injections when making the prior injections

It is also interesting to find that a little shaking by the autosampler arms moving the vials every 40min (the run time of an analysis) can prevent the settlement effectively. Ten continuous injections were made from the same sample vial, showing that the analysis can be repeated nicely (Figure 4).

Based on the above, all the diluted samples were shaken well no more than 40 min before the injection. This practice is very critical when doing particle related analysis of this type of sample.

3.2. Particle Size Distribution Related Information from AFFFF-MALS-DLS-RI Analysis

3.2.1. On-line UV Measurement

A particle size distribution (PSD) is a description of the size and frequency of particles in a population (*17*) that provides an index (means of expression) indicating what sizes (radius or diameter of spherical particle, or equivalent size of non-spherical particle) of particles are present in what proportions (relative particle amount as a percentage where the total amount of particles is 100%) in the sample particle group to be measured. Therefore, both the size and the frequency (usually in terms of % number, % volume or % mass) of the particle with that specific size should be measured to obtain the particle size distribution. For the % mass determination, a UV detector has been widely used as a concentration detector for FFF and can provide precise data. However, there are some disadvantages in using UV as a concentration detector. First, the scattered light from particles can also contribute to the UV detector response. This issue has been recognized for particle characterization using FFF-UV without light scattering correction (*18*). Mie scattering correction has been used to correct the scattering contribution to better determine the concentration (*19*). The light scattering correction for the concentration determination of particles has been incorporated into some commercial software for CHDF (*20*, *21*) , HDC (*22*) and DCP (*23*); however, the version of Astra software currently used at Dow for this study does not have the built in scattering function correction. Second, the UV extinction coefficient of colloid silica particles is very low, and the signal to noise ratio is much worse than other detectors. Figure 6 is the overlaid fractograms from different detectors used in this study. The UV detector, shown using green dots, gave the worst S/N ratio. Therefore, UV data was not used in this study although it can be used to quantify other particles.

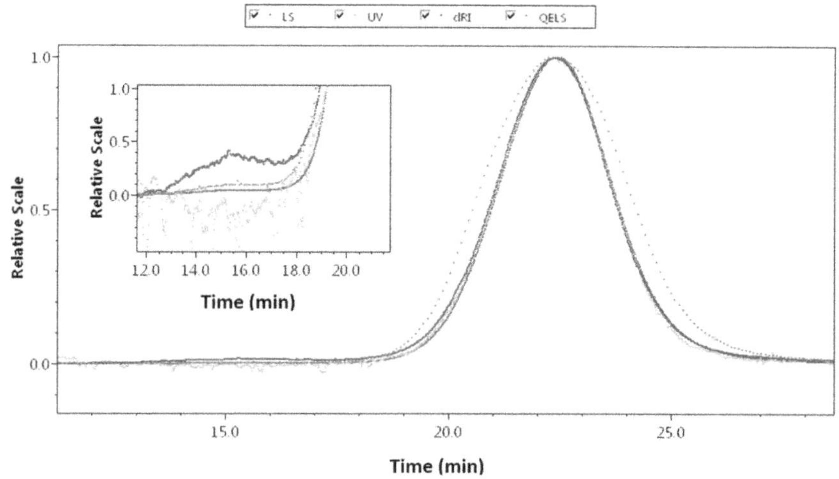

Figure 5. AFFFF fractograms of lot 1 1501-50 sample using different detectors: light scattering at 90 degree angle (LS, in red), UV at 254nm (UV, in green), differential refractive index (dRI, in blue) and Dynamic light scattering (DLS also known as QELS, in magenta).

3.2.2. On-line dRI Measurement and Refractive Index Increment

A differential refractive index detector (dRI) is the most common concentration detector used in Gel Permeation Chromatography (GPC) to determine the sample concentration at each elution moment. However, it is very sensitive to back pressure change, and was not originally used for flow FFF due to the huge response change caused by system pressure changes during the flow FFF separation.. This problem can be mathematically minimized by subtracting the RI fractogram of a blank injection with the assumption that the influence of system pressure by samples is negligible. Meanwhile, the problem can also be minimized by optimizing the AFFFF method for the targeted samples, for example by using a constant or linear cross-flow profile instead of multiple stepwise changes. The obtained dRI signal using the AFFFF method in this study can provide decent baseline stability and good S/N ratio as shown in Figure 5 (shown in dark blue).

At each elution moment, the sample concentration can be determined based on dRI response, instrument calibration constant and the refractive index increment (dn/dc) of the sample in the mobile phase at the measurement wavelength. The dn/dc of a typical 1501-50 sample was determined to be 0.066±0.001 mL/g using offline measurement (*24*). Therefore, the dRI peak area can be converted to the recovered mass of the sample combined with the flow rate used. The ratio of the recovered mass and the injected mass is usually called the mass recovery. The particle size distribution will be reliable only when full mass recovery can be achieved after separation, otherwise the reported distribution only reflect the portion being detected.

It is worth pointing out that there are different opinions in the open literature on the dn/dc dependence on particle size. Theoretically, based on Mie theory, Zimm and Dandliker derived a general refractive index expression (*25*) given by equation 1:

$$\frac{dn}{dc} = \frac{3n}{2\alpha^3 D} R(i_1^*\ ) \qquad \textbf{Equation 1}$$

Where R(i_1*) is obtained from tabulations of the Mie theory, the density of the particle is D and α=2πr/λ where r is sphere radius and λ is the wave length in the medium. For small spheres the refractive index equation reduces to the well-known result:

$$\frac{dn}{dc} = \frac{3n}{2D}\frac{(m^2-1)}{(m^2+2)} \qquad \textbf{Equation 2}$$

where m is the relative refractive index.

Experimentally, measurements by Silebi and McHugh (*26*) show a surprising agreement between measured data and the equation derived by Zimm and Dandliker for polystyrene latex particles as large as 350 nm. However, subsequent data measured by Nagy (*27*) from the same laboratory indicate that dn/dc reverses in sign with increasing particle size, implying that the signal is null for some intermediate particle sizes. Coll and Fague (*28*) observed that dn/dc was independent of concentration for a given latex, thought its value increased linearly with latex particle diameter. As pointed out by Hamielec (*29*), depending on the refractive indexes of particle and dispersion medium, dn/dc may be either increase with particle size or decrease and can change sign.

In recent years, latex and colloid particle characterization has been reported for HDC and GPC using dRI detectors (*10*, *30–33*). The dn/dc of different non-spherical colloid silica particles determined by off-line dn/dc measurement were reported to be 0.079 ±0.007 mL/g (*10*) and 0.083 ±0.002 mL/g (*32*), which are higher than what we measured in this study for Klebosol® 1501-50. Based on the reported refractive index (n) and the specific gravity (D) values of the polymorphs of silica containing silicon in tetrahedral coordination, the n and D of silica glass are 1.46 and 2.2 respectively (*34*). When the small particle limit is satisfied so that2, can be used, the calculated dn/dc will be 0.065 mL/g in the mobile phase (n=1.330), which is surprisingly close to our measured value. The reason that we feel it is surprising is that the specific gravity of the Klebosol® 1501-50 slurry with 30 weight% solid is 1.2. Therefore, the particle specific gravity should be 2.25 if there is no volume change during the mixing of silica particle and water. Based on this measured specific gravity and the reported relationship between refractive index and specific gravity (*34*) as shown in Figure 6, the calculated dn/dc is very close to what we measured off-line. In other words, for the silica particles in this study, the dn/dc can be calculated based on specific gravity and weight% solid of the sample without any concern for the particle size effect. The higher dn/dc reported by Brewer and Striegel (*10, 30–33*) might be due to the larger particle size, non-spherical shape and difference in mobile phase condition.

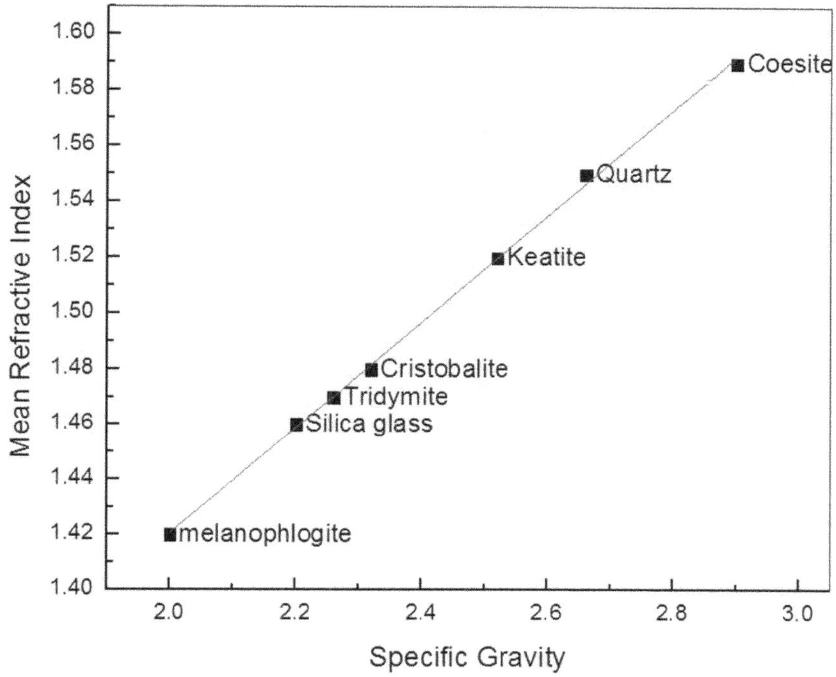

Figure 6. Linear relation between mean refractive index and specific gravity observed for the polymorphs of silica containing silicon in tetrahedral coordination. Data from reference (34) and the references therein.

3.2.3. On-line MALS-dRI Measurement

As demonstrated in Figure 7, the geometric radius (Rgeo) at each elution time is determined based on excess scattering intensity measured by MALS at different angles using a sphere model. The scattering intensities from detector 4 to 15 (the scattering angle range is between 24.9° to 132.2°) were used to calculate the particle size in terms of Rgeo with R^2=0.9996 fit at elution time peak position of 21.969 min.

Similarly, the Rgeo at each elution moment can be obtained across the entire peak as shown by the dots in Figure 8. For most of the data points, the Rgeo has good precision; however, for the small particles eluted before 15.8 min, the Rgeo data are scattered, which is due to low concentration and small size. In order to improve the precision of PSD, Rgeo for these small particles can be estimated by extrapolating the Rgeo vs retention time line at 15.8 to 18 min as shown by the black line before 15 min. By combining Rgeo from fitted data for particles eluted before 15.8 min and the Rgeo measured from MALS for particles eluted after 15.8

min, better precision should be obtained than using all measured Rgeo data. After exporting the dRI and Rgeo data from the Astra file, and reducing the data to have the same step using Origin software, the differential and cumulative PSD plots can be obtained (Figure 9). We also want to point out that the PSD plots using Astra software version 6.1 were not correct; the wt% value of the small mode is very low (<0.5%) and does not agree with the dRI peak integration for the small mode. Although the PSD can be obtained manually, we felt this was not a good approach to quantify the small mode of this bi-modal silica particle samples due to the noise level of the Rgeo for small particles, in addition to the commercial software problems. Therefore, other ways to quantify the small mode and describe the distribution were pursued.

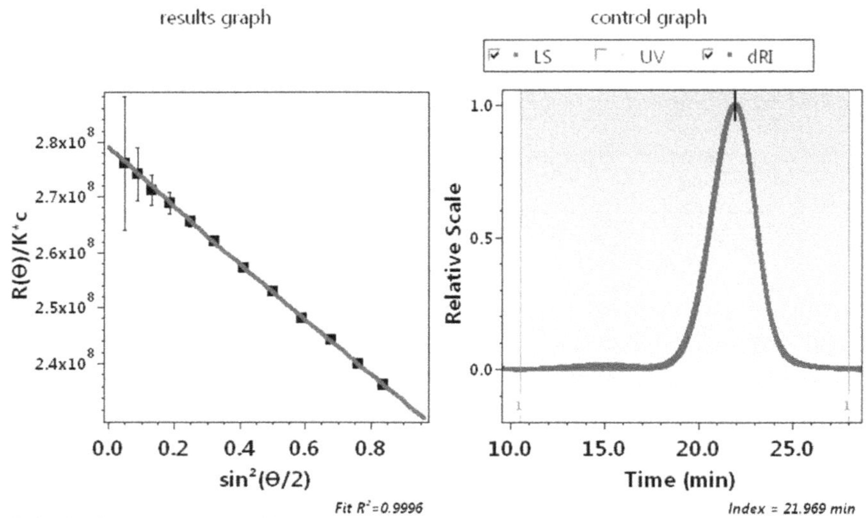

	Value
Molar Mass	(2.792 ± 0.001) e+8 g/mol
radius	38.9 ± 0.3 nm
Peak Number	1
Slice Index	1319
Model	Sphere
Fit Degree	n/a
Abscissa Position	21.969 min
Concentration	(5.065 ± 0.002) e-2 mg/mL
dn/dc (mL/g)	0.0662

Figure 7. AFFFF-MALS-dRI for particle size and particle molar mass determination. Fractograms of Klebosol® 1501-50 lot 1 (right plot, LS response in red and RI response in blue), Debye plot based on on-line MALS measurement at the peak maximum (left plot), and the determined sample concentration, geometric radius and particle molar mass based on sphere model of the fraction eluted at 21.969 min (bottom table).

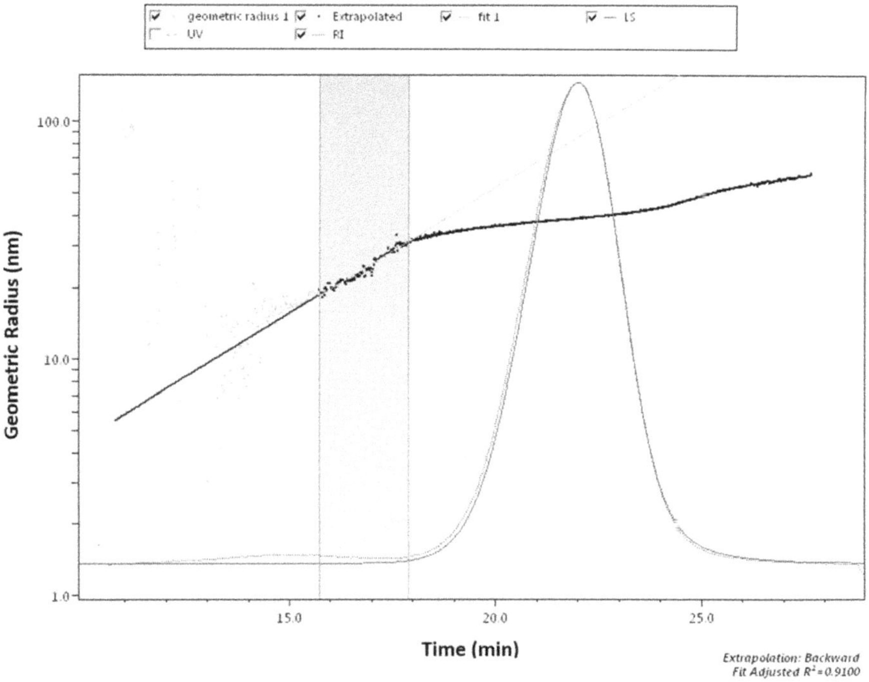

Figure 8. Geometric radius determined from on-line MALS versus AFFFF elution time plot of Klebosol® 1501-50 lot 1. The dots are Rgeo determined from MALS, the solid line before time 15.8 min is the fitted data based on the data in the gray range.

The number density could also be calculated based on the absolute scattering intensity, the instrument physical parameters, mobile phase properties, and sample refractive index value using the patented algorithm (*35*) by Wyatt software. For example, the number density was determined to be 7.4x10^{10} particles/mL at the peak maximum, eluted at 21.969 min (Figure 10). Therefore, the total number, the total volume and the total surface area of the eluted spherical uniform particles in the integrated peak range could be determined theoretically with known flow rate information of the fractionation. In practice, even though the measurement of polystyrene particle standards can be very reasonable, the measurement errors can be significant depending on the sample size range and measurement focus. For small particles, the data can become much noisier especially at very low concentrations as discussed above. The total number and surface area of particles are more sensitive to small particles. Therefore, they can be more sensitive to the small mode population. Meanwhile the measurement noise can also be bigger. In this study we calculated the total number, the total volume and the total surface area.

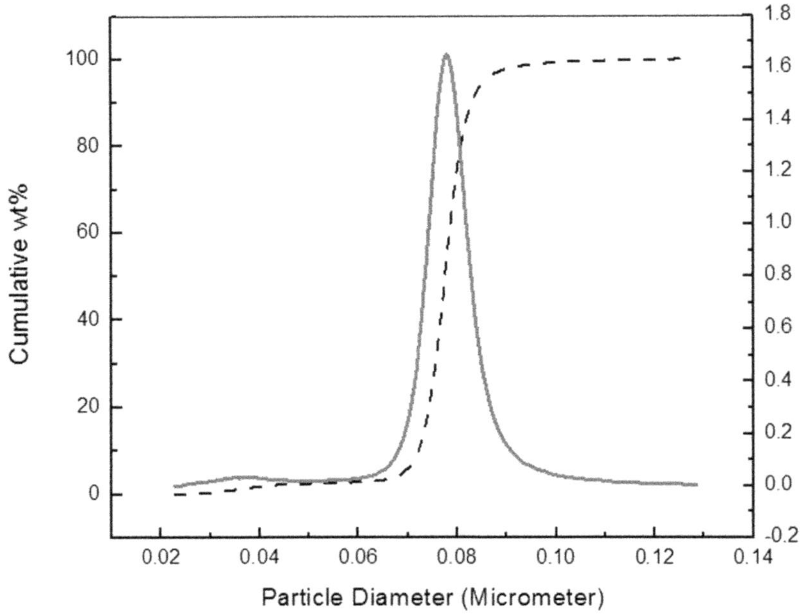

Figure 9. Cumulative (left axis, dotted trace) and differential (right axis, solid trace) particle size distributions of Klebosol® 1501-50 lot 1 obtained from AFFFF-MALS-dRI measurement.

With the concentration determined by the dRI detector and the dn/dc value, the particle molar mass can also be determined. Since the molar mass is from the intercept of the Debye plot, it is not limited by angular dependence of the scattering light and has better precision than the Rgeo measurement as demonstrated in Figure 11. The cumulative and differential molar mass distribution can be obtained using Astra software as shown in Figure 12. The number and weight average molecular weights for the whole peak and the small mode peak were calculated in this study; the small mode peak was of particular interest for understanding whether its variation could lead to variation in removal rate observed from batch to batch.

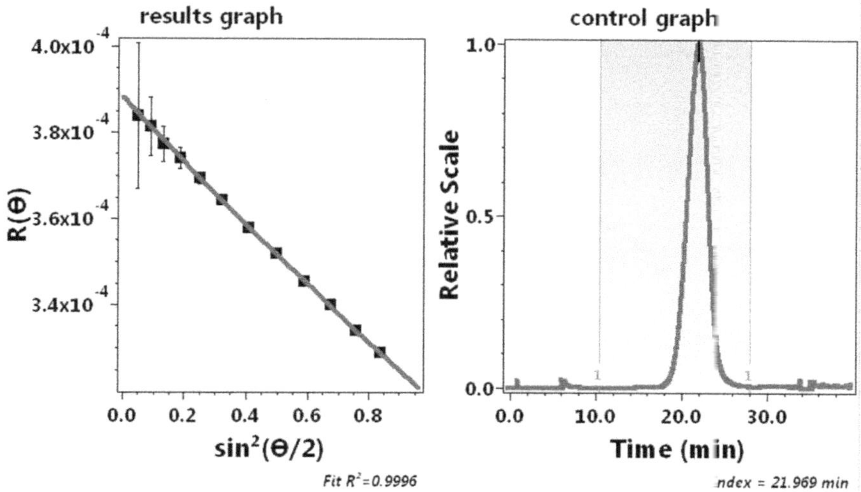

	Value
radius	38.9 ± 0.3 nm
Number Density	(7.4114 ± 0.3134) e+10 particles/mL
Peak Number	1
Slice Index	1319
Model	Sphere
Fit Degree	n/a
Abscissa Position	21.969 min

Figure 10. AFFFF fractograms of Klebosol® 1501-50 lot 1 (right plot), and the Debye plot based on on-line MASL measurement at the peak maximum using hard sphere model (left plot), and the calculated radius and number density of the particles eluted at 21.969 minutes.

It is interesting to point out that the molar mass of the particles eluted at the maximum peak position with Rgeo =38.9 nm was determined to be 2.4x10^8 g/mol. If we assume the Rgeo is the radius of the hard sphere particle, the particle density calculated based on molar mass and Rgeo will be much lower than 2.25. This disagreement might indicate that the 1501-50 silica particles are not as dense as hard spheres made from silica glass. This is not surprising, since they are nucleated spheres made from seed particles and should have a certain porosity associated with them. If the particle morphology is an important factor for applications in the future, more studies could be performed to study the morphology of the particles in the wet state. One of the approaches could be to combine MALS with Dynamic Light Scattering (DLS) to assess the particle density/porosity for spherical particles.

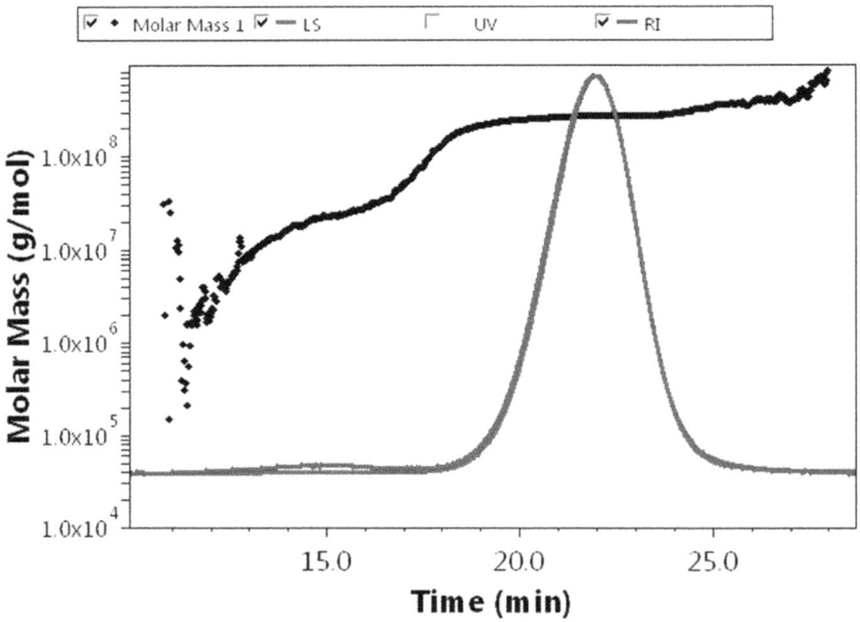

Figure 11. Molar Mass of particles determined from on-line MALS-dRI vs. AFFFF elution time plot of Klebosol® 1501-50 lot 1.

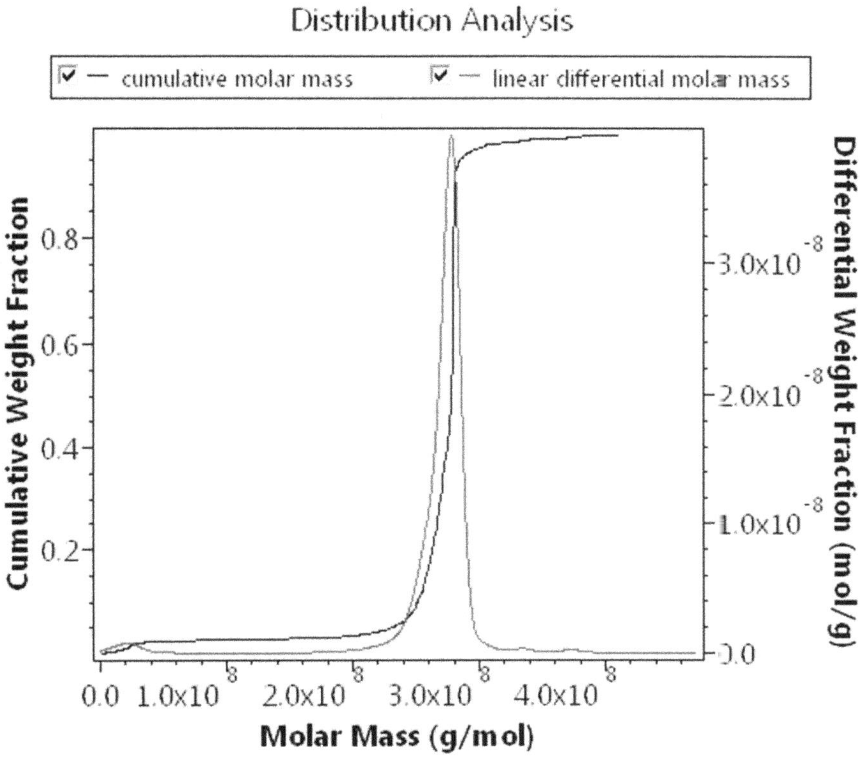

Figure 12. Cumulative (left axis, black) and differential (right axis, red) molar mass distributions of Klebosol® 1501-50 lot 1 obtained from AFFFF-MALS-dRI measurement.

3.2.4. On-line DLS-dRI Measurement

Dynamic light scattering (DLS) is also known as Quasi-Elastic Light Scattering (QELS) or photon correlation spectroscopy (PCS). The Wyatt QELS instrument measures rapid fluctuations in scattered light intensity to determine the translational diffusion coefficient and hydrodynamic radius (R_h) for macromolecules in solution or particles in suspension. The hydrodynamic radius is the radius of an equivalent hard sphere that would feel the same drag force exerted by the flow as is experienced by a macromolecule or particle in solution, and it is defined as:

$$R_h \stackrel{\text{def}}{=} \frac{k_B T}{6\pi\eta_0 D_T} \qquad \textbf{Equation 3}$$

where k_B is Boltzmann's constant, T is the absolute temperature of the solution, η_0 is the viscosity of the solvent, and D_T is the translational diffusion coefficient of the analyte in solution. When the flow rate is low enough and the particle is small enough, R_h at each elution moment can be measured on-line. The measured R_h values of polystyrene (PS) beads of 40 nm and 60 nm particle size standards under the same experimental condition were as expected. The correlation function of a representative sample at peak max using on-line DLS and the obtained R_h and D_T are shown in Figure 13 and Figure 14, however, they became scattered at the small mode peak. Therefore, it will not be as precise as using molar mass to characterize the small mode.

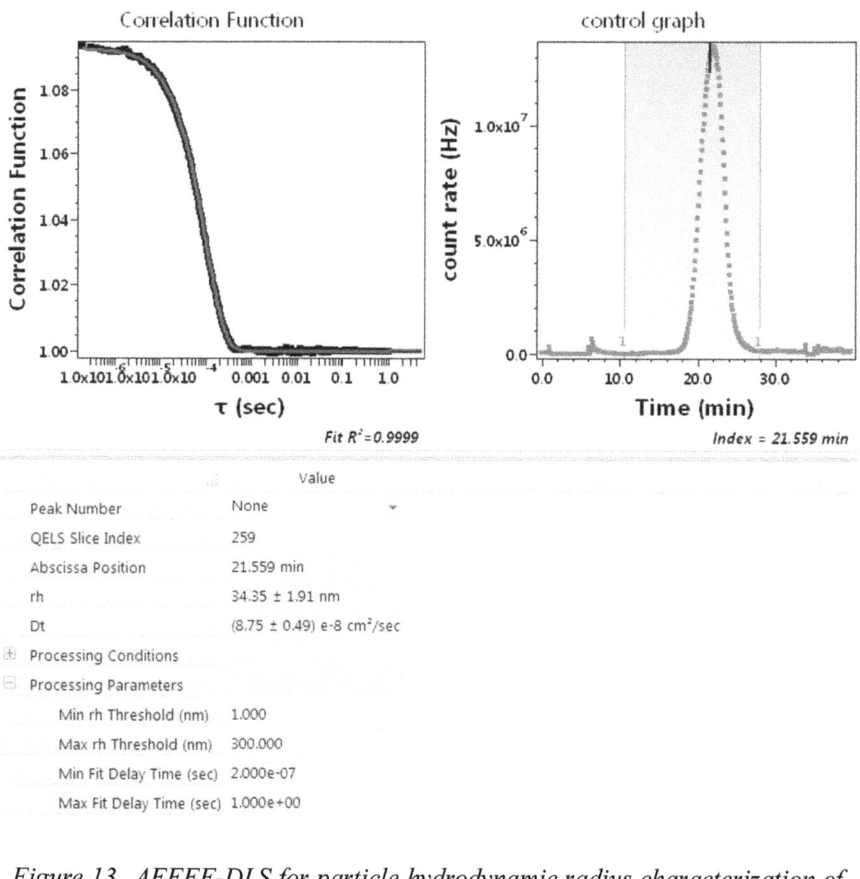

Figure 13. AFFFF-DLS for particle hydrodynamic radius characterization of Klebosol® 1501-50 lot 1. The hydrodynamic radius and diffusion constant of the fraction eluted at peak max (right plot) were determined based on the obtained correlation function (left plot).

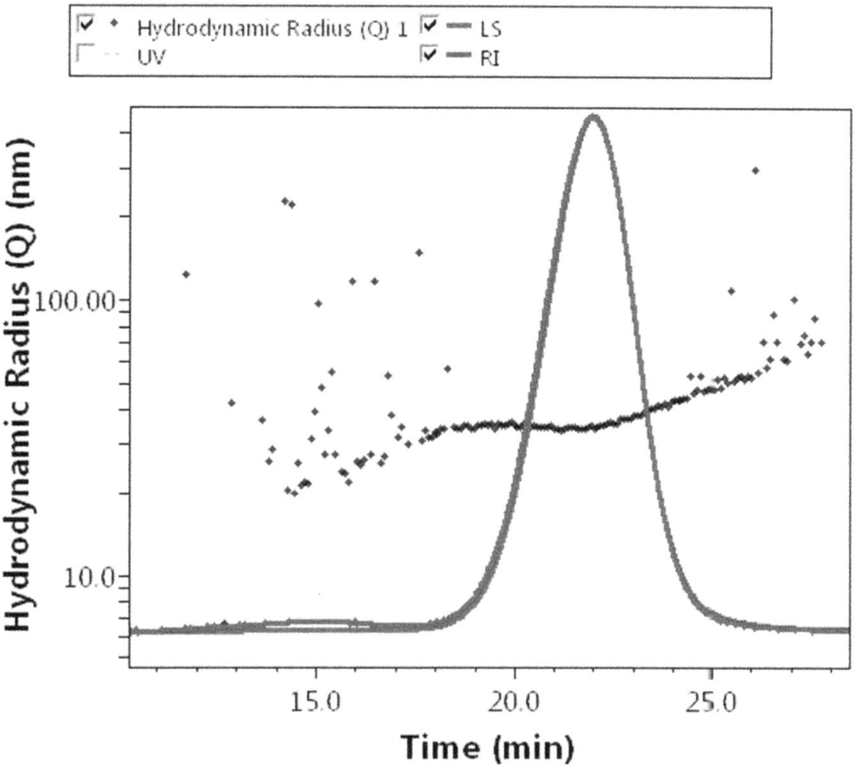

Figure 14. Hydrodynamic radius of particles determined from on-line DSL versus AFFFF elution time of Klebosol® 1501-50 lot 1.

3.2.5. AFFFF-MALS- DLS-dRI Measurement

As described in the previous sections, sample concentrations and size related information can be obtained from AFFFF-MALS-DLS-dRI measurements. This section compares the pros and cons of using different detectors for 1501-50 analysis, especially the precision of the small mode characterization. The related information from AFFFF-MALS-DLS-dRI measurements discussed in this study is listed in Table 3. The property related to each term and relative precision based on the instruments used are also listed to help statistical analysis, since information that comes from the same instrument and describes similar properties in a different form might be highly correlated. Many of these properties were also calculated on different integration ranges, such as for the whole peak (peak 1 eluted from 10.5 to 28 min labeled on the upper right plot of Figure 13) and for the small mode peak (peak 2 eluted from 10.5 to 17.5 min labeled on the upper right plot in Figure 13).

Table 3. Indices from AFFFF-MALS-DLS-dRI Measurement

Symbol	*Name*	*Property*	*Detector instrument*	*Factors affectingprecision*
LS peak area	Peak area of 90 degree light scattering response of the integrated peak	Combined size, PSD and conc.	MALS	Instrument condition, Concentration
LS peak height	Peak height of 90 degree light scattering peak	Combined size, and concentration at peak max	MALS	Instrument condition, Concentration, flow rate control
Small mode LS ratio	The LS peak area ratio of small mode to whole peak	Size and small mode%	MALS	Flow rate control
Rgeo	Geometric radius at each elution moment	size	MALS	Instrument condition, Conc.
Rg	Radius of gyration at each elution moment	size	MALS	Instrument condition, Conc., data fitting model
N	Total number of particles eluted in the integrated peak	combined	MALS	Instrument condition, Conc. and separation
V	Total volume of particles eluted in the integrated peak	combined	MALS	Instrument condition, Conc. and separation
S	Total surface area of particles eluted in the integrated peak	combined	MALS	Instrument condition, Conc. and separation
RI peak area	Peak area of differential refractive index detector response of the integrated peak	conc.	dRI	Instrument condition, Conc., flow rate control
RI peak height	Peak height of dRI peak	Concentration at peak max	dRI	Instrument condition, Conc., flow rate control

Symbol	*Name*	*Property*	*Detector instrument*	*Factors affectingprecision*
Small mode RI ratio	The RI peak area ratio of small mode to whole peak	Small mode %	dRI	Flow rate control
M	Molar Mass of the particle at each elution moment	Size (mass)	MALS-dRI	Instrument condition, Conc., data fitting model
Rh(Q)	Hydrodynamic radius at each elution moment	size	DLS	Instrument condition, background noise, flow rate
Dt	Diffusion constant at each elution moment	size	DLS	Instrument condition, background noise, flow rate
Rn geo, Rw geo, Rz geo	Number, weight, and, z-average of Rgeo	PSD	MALS or MALS-dRI	Instrument condition, Conc. and separation
Mn, Mw, Mz, M(avg), Mp	Number, weight, z, and uncertainty-weighted averages of M and the M at peak	MWD	MALS-dRI	Instrument condition, Conc. and separation
Polydispersity (Mw/Mn)	Polydispersity index in forms of Mw/Mn	MWD	MALS-dRI	Instrument condition, Conc. and separation
Rn, Rw, Rz, R(avg)	Number, weight, z, and uncertainty-weighted averages of Rg	PSD	MALS-dRI	Instrument condition, Conc. and separation
Polydispersity (Rw/Rn)	Polydispersity index in forms of Rw/Rn	PSD	MALS-dRI	Instrument condition, Conc. and separation
Rh(Q)n, Rh(Q)w, Rh(Q)z, Rh(Q)(avg),	Number, weight, z, and uncertainty-weighted averages of hydrodynamic radius	PSD	DLS-dRI	Instrument condition, Conc. and separation
Dt(n), Dt(w), Dt(z), Dt(avg)	Number, weight, z, and uncertainty averages of Number average of diffusion constant	PSD	DLS-dRI	Instrument condition, Conc. and separation

3.3. Evaluation of Lot-to-Lot Variation of Klebosol® 1501-50 Using AFFFF-MALS-DLS-dRI

The relative standard deviations of multiple injections from the control lot (Klebosol® 1501-50 lot 1) obtained from AFFFF-MALS-DLS-dRI measurement are listed as RSD_m (relative standard deviation of multiple injections of control sample) in Table 4 and Table 5. RSD_m lower than 1% can be achieved for most of the indices for the whole sample. The small mode population in Klebosol® 1501-50 can also be separated and quantified with test RSD lower than 10%.

15 lots of Klebosol® 1501-50 were analyzed using the AFFFF-MALS-DLS-dRI technology described above. The lot numbers and obtained results are listed in Table 5. About 30 different indices were obtained from the analysis. Percentage measurement variation (%MV) is used to evaluate measurement system capabilities of each index. It is calculated with the following equation:

$$\%MV = \frac{SD_m}{SD_T} \times 100\% = \frac{SD_m}{\sqrt{SD_m^2 + SD_{ltl}^2}} \times 100\%$$

in which SD_m is the measurement standard deviation, SD_T is the total standard deviation, and SD_{ltl} is the lot-to-lot variation. The SD_m were obtained from multiple injections of the same sample (control lot 1), and the standard deviation of 15 lots were used as SD_{ltl}. The average values of the indices from 15 lots, SD_{ltl}, SD_m, relative measurement standard deviation (RSD_m) of multiple injections from the control lot, and the calculated %MV values for each index are listed in Table 4 and Table 5 for whole peak and small mode peak, respectively.

The criteria for assessment of %MV is (*36, 37*):

%MV < 10%	Excellent
10% < %MV <=20%	Good
20% < %MV <=30%	Marginal
%MV > 30%	Unacceptable

A low %MV (<30%) means the measurement system is acceptable, and can be used as a basis of decision making. The indices in Table 4 and Table 5 are sorted by %MV. Based on above criteria, the AFFFF technology can provide seven excellent measurements (%MV<10%) for the whole peak (Table 4) of this set of 1501-50 lots; these measurement indices are Mp, M(avg), R(avg), LS peak height Rz, Mz and LS peak area. In other words, since the measurement variation is small, the measurement system has excellent capability in detecting sample-to-sample variation. Because the samples are from different lots, the measurement system can compare lot-to-lot differences with great confidence. In addition, there are also a group of five obtained measurement indices providing good (10% < %MV <=20%) capability in detecting lot-to-lot differences; these includes Mw, Rz geo, RI peak area, Rw and volume of particles data. There are measurements falling into the unacceptable range (%MV>30%), for example, the polydispersity of the whole sample peak, which should not be used to compare lot-to-lot differences.

For the small mode characterization (Table 5), there are six measurements that are excellent, and seven good measurements for detecting lot-to-lot differences. These thirteen measurements are small mode LS ratio, LS peak area, Mz, Rz, polydispersity (Mw/Mn), LS peak height, Dt (avg), and RI peak height calculated mass, mass recovery, RI peak area, Mass fraction, small mode RI ratio. Therefore, the AFFFF-MALS-DLS-RI technique is capable of detecting lot-to-lot differences in the small mode with these indices.

Among the above measurements, the averages of particle molar mass, particle size and the small mode level in terms of small mode LS peak area % and RI peak area % of these 15 lots are identified as the key analytical results for these colloidal silica raw abrasive particles (Table 6).

Table 4. Average and Standard Deviation of 15 Lots of 1501-50(SD_{ltl}), Standard Deviation (SD_m) and Relative Standard Deviation (RSD_m) of Multiple Injections of Control Lot, and Percentage Measurement Variation (%MV) of Measured Values Using AFFFF-MALS-DLS-dRI Based on the Whole Peaks of 15 Lots of 1501-50 Silica Slurries

Index	*Average of 15 lots of 1501-50*	*SD_{ltl} (15 lots of 1501-50)*	*SD_m of multiple inj. of control lot*	*RSD_m of multiple inj. of control lot*	*%MV*
Mp (kDa)	2.55E+05	8.67E+03	4.54E+02	0.18%	5%
M(avg) (kDa)	2.51E+05	8.34E+03	5.25E+02	0.21%	6%
R(avg) (nm)	2.81E+01	2.58E-01	1.81E-02	0.06%	7%
LS peak height (1/cm)	2.95E-04	1.72E-05	1.40E-06	0.49%	8%
Rz (nm)	2.83E+01	2.83E-01	2.48E-02	0.09%	9%
Mz (kDa)	2.53E+05	9.14E+03	9.03E+02	0.37%	10%
LS peak area (1/cm min)	8.83E-04	4.61E-05	4.62E-06	0.54%	10%
Mw (kDa)	2.39E+05	1.09E+04	1.39E+03	0.59%	13%
Rz geo (nm)	3.76E+01	3.86E-01	5.35E-02	0.14%	14%
RI peak height (RIU)	3.04E-06	9.22E-08	1.64E-08	0.54%	17%
Rw (nm)	2.77E+01	3.88E-01	7.79E-02	0.28%	20%
Volume of particles (μL)	4.89E-02	1.39E-03	2.82E-04	0.58%	20%
Rh(Q)w (nm)	3.50E+01	4.41E-01	9.50E-02	0.27%	21%
Rh(Q)(avg) (nm)	3.56E+01	2.26E+00	5.74E-01	1.54%	25%
Total number of particles	2.22E+11	5.97E+09	1.60E+09	0.72%	26%

Continued on next page.

Table 4. (Continued). Average and Standard Deviation of 15 Lots of 1501-50(SD_{ltl}), Standard Deviation (SD_m) and Relative Standard Deviation (RSD_m) of Multiple Injections of Control Lot, and Percentage Measurement Variation (%MV) of Measured Values Using AFFFF-MALS-DLS-dRI Based on the Whole Peaks of 15 Lots of 1501-50 Silica Slurries

Index	*Average of 15 lots of 1501-50*	*SD_{ltl} (15 lots of 1501-50)*	*SD_m of multiple inj. of control lot*	*RSD_m of multiple inj. of control lot*	*%MV*
Surface area of particles (m^2)	3.91E-03	9.12E-05	2.48E-05	0.64%	26%
Calculated mass (μg)	1.49E+02	2.96E+00	8.16E-01	0.56%	27%
RI peak area (RIU min)	9.81E-06	1.95E-07	5.41E-08	0.57%	27%
Mass recovery (%)	9.91E+01	1.97E+00	5.49E-01	0.57%	27%
Rh(Q)n (nm)	3.46E+01	3.20E+00	1.17E+00	3.25%	34%
UV peak area (AU min)	1.50E-02	9.23E-03	4.43E-03	22.29%	43%
Dt(w) (cm^2/sec)	8.78E-08	1.39E-09	5.09E-10	0.58%	35%
Rh(Q)z (nm)	3.54E+01	2.59E-01	1.00E-01	0.29%	36%
Dt(n) (cm^2/sec)	9.95E-08	6.23E-09	2.42E-09	2.57%	36%
Dt(avg) (cm^2/sec)	5.31E-08	9.05E-09	4.45E-09	7.38%	44%
Mn (kDa)	1.65E+05	1.88E+04	1.20E+04	6.89%	54%
Dt(z) (cm^2/sec)	8.59E-08	7.06E-10	5.03E-10	0.58%	58%
Polydispersity (Mw/Mn)	1.46E+00	1.13E-01	8.97E-02	6.56%	62%
Rn (nm)	2.40E+01	8.78E-01	9.86E-01	4.09%	75%

Table 5. Average and Standard Deviation of 15 Lots of 1501-50(SD_{ltl}), Standard Deviation (SD_m) and Relative Standard Deviation (RSD_m) of Multiple Injections of Control Lot, and Percentage Measurement Variation (%MV) of Measured Values Using AFFFF-MALS-DLS-dRI Based on the Small Mode Peaks of 15 Lots of 1501-50 Silica Slurries

Index	*Average of 15 lots of 1501-50*	*SD_{ltl} (15 lots of 1501-50)*	*SD_m of multiple injections of control lot*	*RSD_m of 10 inj. of contr-ol lot*	*%MV*
small mode LS ratio	4.93E-03	2.29E-03	2.90E-05	1.4%	1%
LS peak area (1/cm min)	4.27E-06	1.80E-06	2.61E-08	1.5%	1%
Mz (kDa)	4.88E+04	9.27E+04	2.18E+03	12.1%	2%
Rz (nm)	3.08E+01	5.67E+01	2.42E+00	18.8%	4%
Polydispersity (Mw/Mn)	1.15E+00	1.93E-01	8.42E-03	0.8%	4%
LS peak height (1/cm)	1.55E-06	4.49E-07	2.12E-08	2.7%	5%
Dt(avg) (cm^2/sec)	4.57E-08	2.87E-08	3.00E-09	12.8%	10%
RI peak height (RIU)	1.35E-07	4.79E-08	5.14E-09	6.6%	11%
Calculated mass (μg)	7.73E+00	2.33E+00	3.54E-01	7.0%	15%
Mass recovery (%)	5.16E+00	1.56E+00	2.36E-01	7.0%	15%
RI peak area (RIU min)	5.10E-07	1.54E-07	2.34E-08	7.0%	15%
Mass fraction (%)	5.21E+00	1.58E+00	2.42E-01	6.9%	15%
small mode RI ratio	5.21E-02	1.58E-02	2.42E-03	6.9%	15%
M(avg) (kDa)	1.82E+04	6.49E+03	1.58E+03	12.2%	24%
Rw (nm)	1.80E+01	1.06E+01	2.61E+00	22.2%	24%
Mp (kDa)	1.94E+04	2.84E+03	9.33E+02	7.5%	31%
Mw (kDa)	2.33E+04	4.58E+03	1.86E+03	11.5%	38%
Rh(Q)z (nm)	3.35E+01	9.13E+00	4.07E+00	11.0%	41%
Rh(Q)w (nm)	3.39E+01	9.65E+00	4.33E+00	11.5%	41%
Rh(Q)n (nm)	3.44E+01	1.00E+01	5.07E+00	13.2%	45%
Dt(z) (cm^2/sec)	1.21E-07	1.11E-08	6.18E-09	5.5%	49%
Dt(w) (cm^2/sec)	1.23E-07	1.10E-08	7.01E-09	6.1%	54%
Dt(n) (cm^2/sec)	1.25E-07	1.10E-08	7.53E-09	6.5%	56%
UV peak height (AU)	3.13E-04	3.25E-04	2.39E-04	68.5%	59%
Mn (kDa)	2.03E+04	1.84E+03	1.65E+03	10.9%	67%
R(avg) (nm)	1.79E+01	2.62E+00	2.81E+00	19.6%	73%

Continued on next page.

Table 5. (Continued). Average and Standard Deviation of 15 Lots of 1501-50(SD_{ltl}), Standard Deviation (SD_m) and Relative Standard Deviation (RSD_m) of Multiple Injections of Control Lot, and Percentage Measurement Variation (%MV) of Measured Values Using AFFFF-MALS-DLS-dRI Based on the Small Mode Peaks of 15 Lots of 1501-50 Silica Slurries

Index	*Average of 15 lots of 1501-50*	*SD_{ltl} (15 lots of 1501-50)*	*SD_m of multiple injections of control lot*	*RSD_m of 10 inj. of contr-ol lot*	*%MV*
Rn (nm)	1.48E+01	1.97E+00	2.81E+00	25.8%	82%
Rh(Q)(avg) (nm)	2.27E+01	3.46E+00	6.06E+00	27.2%	87%

Table 6. Key Analytical Results on 15 lots of Klebosol® 1501-50 Colloidal Silica Raw Abrasive Particles Obtained from AFFFF-MALS-DLS-dRI Measurement

Lot	1	2	3	4	5	6	7	8	9	10	11	12	13	14	15	Ave	RSD
Mp (kDa) $x10^{-3}$	255	262	270	268	260	254	239	253	254	244	255	256	246	255	251	255	3%
M(avg) (MDa) $x10^{-3}$	251	257	265	263	255	248	234	248	250	239	251	252	242	250	247	251	3%
R(avg) (nm)	28	28.4	28.6	28.5	28.3	28.0	27.7	28.1	28.1	27.8	28.2	28.2	27.9	28.1	28.0	28.1	1%
Rz (nm)	28	28.6	28.8	28.8	28.5	28.1	27.8	28.2	28.3	28.0	28.4	28.3	28.2	28.2	28.2	28.3	1%
Mz (kDa) $x10^{-3}$	253	259	272	267	260	250	236	248	251	242	252	252	248	251	248	253	4%
Mw (kDa) $x10^{-3}$	239	248	257	255	248	236	218	229	240	227	240	239	232	235	236	239	5%
Rz geo (nm)	38	37.9	38.2	38.1	37.7	37.4	36.8	37.6	37.5	37.0	37.6	37.7	37.2	37.6	37.5	37.6	1%
Rw (nm)	28	28.0	28.2	28.2	27.9	27.5	26.8	27.4	27.8	27.3	27.8	27.7	27.5	27.5	27.7	27.7	1%
small mode LS area %	0.5	0.32	0.29	0.31	0.24	0.48	1.00	0.90	0.40	0.67	0.43	0.51	0.52	0.67	0.41	0.49	46%
small mode RI area %	5.2	3.99	3.77	3.69	3.65	5.20	8.31	8.40	4.30	5.88	4.86	5.59	5.36	6.72	4.93	5.21	30%

4. Conclusion

Asymmetrical Flow Field Flow Fractionation with online multi-angle light scattering, dynamic light scattering and differential refractive index detectors (AFFFF-MALS-DLS-dRI) can measure the particle size, concentration and size distribution of colloidal silica particle, such as Klebosol® 1501-50. The concentration information in terms of sample weight%, number of particles, surface area of particles and volume of particles can be obtained from MALS and dRI detectors at each fraction separated by AFFFF. The geometric radius (Rgeo), radius of gyration (Rg), hydrodynamic radius (Rh), diffusion constant (Dt) and particle molar mass (MM) can also be determined at each elution moment using on-line detectors. Therefore the particle size and molar mass distribution plots and number, weight, z-averages and uncertainty-weight average of different sizes of the whole peak and the small mode peak can be obtained within a single one hour run using the developed AFFFF-MALS-DLS-dRI method for Klebosol® 1501-50 slurry. The detailed data process and meaning of about thirty indices were explained in this report. With careful sample preparation and injection procedure, lower than 1% relative standard deviation (RSD) of multiple injections of same sample can be achieved for most of these indices for the whole sample. The small mode population in Klebosol® 1501-50 can also be separated and quantified with RSD lower than 10%. Fifteen lots of Klebosol® 1501-50 were analyzed using this developed experimental and data process approach. The lot-to-lot variations of Klebosol® 1501-50 of the whole sample and small mode can be differentiated using more than ten indices with excellent to good acceptance assessed by percentage measurement variation. While information from MALS has the best precision, dRI also can provide precise concentration measurements under the test conditions. The Rh and Dt from DLS are less precise than size information from MALS. Based on dRI measurements, the average level of the small mode are determined to be around 5% with the percentage measurement variation of 15%. Therefore, AFFFF-MALS-dRI is capable of detecting lot-to-lot differences of a set of 15 lots Klebosol® 1501-50 in terms of concentration, particle molar mass and size averages for the whole sample and level of the small mode, etc. This developed AFFFF-MALS-dRI technology is expected to become a powerful tool for the characterization of abrasive particles in the CMP industry for better quality control as well as for fundamental studies to better understand if and how particle size and the particle size distribution can impact removal rate after being formulated into different slurry formulations.

LIST OF ABBREVIATES

Symbol	Name
AFFFF, A4F, AF4	Asymmetrical flow field flow fractionation
CHDF	Capillary hydrodynamic hractionation
DCP	Disk centrifuge photosedimentometry
DLS	Dynamic light scattering (detector)
dRI	Differential refractive index (detector)
Dt	Diffusion constant
Dt(avg)	Uncertainty-weighted average diffusion constant
Dt(n)	Number average diffusion constant
Dt(w)	Weight average diffusion constant
Dt(z)	Z-average diffusion constant
FFF	Field flow fractionation
HDC	Hydrodynamic chromatography
LS	Light scattering
MALS	Multi angle light scattering
MM	Molar mass
M(avg)	Uncertainty-weighted average molar mass
Mn	Number average molecular weight
Mp	Peak molecular weight
Mw	Weight average molecular weight
MW	Molecular weight
Mw/Mn	Polydispersity index in forms of Mw/Mn
Mz	Z-average molecule weight
N	Total number of particles
QELS	Quasi-elastic light scattering
Rg	Radius of gyration
R(avg)	Uncertainty-weighted radius of gyration
Rgeo	Geometric radius
Rh, Rh(Q)	Hydrodynamic radius from DLS/QELS
Rh(Q)(avg)	Uncertainty-weighted hydrodynamic radius
Rh(Q)n	Number average hydrodynamic radius
Rh(Q)w	Weight average hydrodynamic radius
Rh(Q)z	Z-average hydrodynamic radius
RI	Refractive index
Rn	Number average of Rg
Rw	Weight average of Rg
Rw/Rn	Polydispersity index in forms of Rw/Rn
Rz	Z-average of Rg
Rn geo	Number average of Rgeo
Rw geo	Weight average of Rgeo
Rz geo	Z-average of Rgeo
S	Total surface area of particles
UV	Ultraviolet/visible spectroscopic (detector)
V	Total volume of particles

Acknowledgments

We thank Ms. Lindsay Koukoski for the sample preparation for AF4 analysis, Mr. Casey Wolf for TEM image, and Dow Electonic Materials CMP business for silica samples and COA information.

References

1. Klebosol® is a registered trademark of Merck KGaA, D., Germany.
2. Preston, F. The Structure of Abraded Glass Surfaces. *Trans. Opt. Soc.* **1922**, *23*, 141–6.
3. Brown, N. J. A Short Course in Optical Fabrication Technology. *Optical Society of America Annual Meeting* 1981, *LLNL Preprint*, UCRL-86796.
4. Basim, G. B.; Adler, J. J.; Mahajan, U.; Singh, R. K.; Moudgil, B. M. Effect of Particle Size of Chemical Mechanical Polishing Slurries for Enhanced Polishing with Minimal Defects. *J. Electrochem. Soc.* **2000**, *147*, 3523–3528.
5. Suratwala, T.; Feit, M.; Steele, W.; Wong, L.; Shen, N.; Dylla-Spears, R.; Desjardin, R.; Mason, D.; Geraghty, P.; Miller, P.; Baxamusa, S Microscopic Removal Function and the Relationship Between Slurry Particle Size Distribution and Workpiece Roughness During Pad Polishing. *J. Am. Ceram. Soc.* **2014**, *97*, 81–91.
6. Cook, L. Chemical Processes in Glass Polishing. *J. Non-Cryst. Solids* **1990**, *120*, 152–71.
7. Chandrasekaran, N.; Taylor, T.; Sabde, G. Effect of Ceria Particle-Size Distribution and Pressure Interactions in Chemo-Mechanical Polishing (CMP) of Dielectric Materials. *MRS Online Proc. Libr. Arch.* **2003**, *767*.
8. Lu, Z.; Lee, S.-H.; Babu, S. V.; Matijević, E. The use of monodispersed colloids in the polishing of copper and tantalum. *J. Colloid Interface Sci.* **2003**, *261*, 55–64.
9. Bass, J. Particle Size and Characterization Techniques. In *Colloidal Silica Fundamentals and Applications*; Bergna, H. E., Roberts, W. O., Eds.; CRC Press, Talyor & Francis Group: Boca Raton, FL, 2006; p 535.
10. Brewer, A. K.; Striegel, A. M. Characterizing String-of-Pearls Colloidal Silica by Multidetector Hydrodynamic Chromatography and Comparison to Multidetector Size-Exclusion Chromatography, Off-Line Multiangle Static Light Scattering, and Transmission Electron Microscopy. *Anal. Chem.* **2011**, *83*, 3068–3075.
11. Giddings, J. C. Measuring colloidal and macromolecular properties by FFF. *Anal. Chem.* **1995**, *67*, 592A–598A.
12. Janca, J. *Field-flow Fractionation Analysis of Macromolecules and Particles*; M. Dekker Inc, 1988.
13. Schimpf, M. E.; Giddings, J. C.; Caldwell, K. *Field-Flow Fractionation Handbook*; 2000.
14. Wahlund, K. G.; Giddings, J. C. Properties of an asymmetrical flow field-flow fractionation channel having one permeable wall. *Anal. Chem.* **1987**, *59*, 1332–1339.

15. Runyon, J. R.; Ulmius, M.; Nilsson, L. A perspective on the characterization of colloids and macromolecules using asymmetrical flow field-flow fractionation. *Colloids Surf. A* **2014**, *442*, 25–33.
16. Wahlund, K. G.; Nilsson, L. Flow FFF – basics and key applications. In *Field-flow Fractionation in Biopolymer Analysis*; Williams, S. K. R., Caldwell, K. D., Eds;. Springer Verlag: Wien, 2012; pp 1−21.
17. Jillavenkatesa, A.; Lum, L.-S. H.; Dapkunas, S. NIST Recommended Practice Guide: Particle Size Characterization. In *Special Publication (NIST SP) - 960-1*; 2001.
18. Ratanathanawongs, S. K.; Lee, I.; Giddings, J. C. Separation and Characterization of 0.01-50 um Particles using Flow Field-flow Fractionation. In *Particle Size Distribution II Assessment and Characterization*; Provder, T., Ed.; American Chemcial Society: Washington, DC, 1991.
19. Kirkland, J. J.; Rementer, S. W.; Yau, W. W. Time-delayed exponential field-programmed sedimentation field flow fractionation for particle-size-distribution analyses. *Anal. Chem.* **1981**, *53*, 1730–1736.
20. DosRamos, J. G.; Silebi, C. A. The determination of particle size distribution of submicrometer particles by capillary hydrodynamic fractionation (CHDF). *J. Colloid Interface Sci.* **1990**, *135*, 165–177.
21. Miller, C. M.; Sudol, E. D.; Silebi, C. A.; El-Aasser, M. S. Capillary Hydrodynamic Fractionation (CHDF) as a Tool for Monitoring the Evolution of the Particle Size Distribution during Miniemulsion Polymerization. *J. Colloid Interface Sci.* **1995**, *172*, 249–256.
22. Small, H.; Langhorst, M. A. Hydrodynamic Chromatography. *Anal. Chem.* **1982**, *54*, 892A–898A.
23. Kamiti, M.; Boldridge, D.; Ndoping, L. M.; Remsen, E. E. Simultaneous Absolute Determination of Particle Size and Effective Density of Submicron Colloids by Disc Centrifuge Photosedimentometry. *Anal. Chem.* **2012**, *84*, 10526–10530.
24. Gao, W.; Liu, X. M.; Gross, R. A. Determination of molar mass and solution properties of cationic hydroxyethyl cellulose derivatives by multi-angle laser light scattering with simultaneous refractive index detection. *Polym. Int* **.2009**, *58*, 1115–1119.
25. Zimm, B. H.; Dandliker, W. B. Theory of Light Scattering and Refractive Index of Solutions of Large Colloidal Particles. *J. Phys. Chem.* **1954**, *58*, 644–648.
26. Silebi, C. A.; McHugh, A. J. The determination of particle size distribution by hydrodynamic chromatography. An analysis of dispersion and methods for improved signal resolution. *J. Appl. Polym. Sci.* **1979**, *23*, 1699–1721.
27. Nagy, D. J. *Column Chromatography of Polymer Latexes*; Lehigh University, 1979.
28. Coll, H.; Fague, G. R. Liquid exclusion chromatography of colloidal dispersions. *J. Colloid Interface Sci.* **1980**, *76*, 116–123.
29. Hamielec, A. E., Detection Systems for Particle Chromatography. In *Moden Methods of Particle Size Analysis*; Barth, H., Ed.; John Wiley & Sons: New York, 1984.

30. Brewer, A. K.; Striegel, A. M. Particle size characterization by quadruple-detector hydrodynamic chromatography. *Anal. Bioanal. Chem.* **2009**, *393*, 295–302.
31. Brewer, A. K.; Striegel, A. M. Hydrodynamic chromatography of latex blends. *J. Sep. Sci.* **2010**, *33*, 3555–3563.
32. Brewer, A. K.; Striegel, A. M. Characterizing the size, shape, and compactness of a polydisperse prolate ellipsoidal particle via quadruple-detector hydrodynamic chromatography. *Analyst (Cambridge, U.K.)* **2011**, *136*, 515–519.
33. Brewer, A. K.; Striegel, A. M. Characterizing a spheroidal nanocage drug delivery vesicle using multi-detector hydrodynamic chromatography. *Anal. Bioanal. Chem.* **2011**, *399*, 1507–1514.
34. Skinner, B. J.; Appleman, D. E. Melanophlogite, A Cubic Polymorph of Silica. *Am. Mineral.* **1963**, *48*, 14.
35. Wyatt, P. J.; Weida, M. J. *Method and apparatus for determining absolute number densities of particles in suspension*; US 6,774,994 B1, 2004.
36. Sheehy, P.; Navarro, D.; Silvers, R.; Keyes, V.; Dixon, D. *The Black Belt Memory Jogger; A Pocket Guide for Six Sigma Success*, 1st ed.; Goal/QPC: 2002.
37. Barrentine, L. B. *Concepts for R&R Studies*, 2nd ed.; ASQ Quality Press: Milwaukee, Wisconsin: 2003.

Chapter 8

Characterization of Biopolymer Stability by Size-Exclusion Chromatography and Asymmetric Flow Field-Flow Fractionation

Yongfu Li,*,[1] Krishna Kuppannan,[2] and David M. Meunier[1]

[1]The Dow Chemical Company, Core R&D Analytical Sciences, Midland, Michigan 48674, United States
[2]The Dow Chemical Company, Dow AgroSciences LLC, 9330 Zionsville Road, Indianapolis, Indiana 46268, United States
***E-mail: ylli@dow.com.**

Biopolymers in pharmaceutical applications need careful formulation to maintain bioactivity of the drug. Aggregation is one of the common denaturation pathways of biopolymers, and the aggregates might also cause serious undesired biological reaction. Monitoring molecular weight changes can enable development of novel additives to improve stability of biopolymers in formulations. Size-exclusion chromatography (SEC) and asymmetric flow field-flow fractionation (AF4) are widely employed techniques for characterization of biopolymers and their aggregation. In this study, SEC and AF4 method development for characterization of biopolymers, including abatacept, immunoglobulin G (IgG) and bovine serum albumin (BSA), in their native states and after heat stress will be discussed. Detailed comparison of SEC versus AF4 will be presented. The results suggest that SEC and AF4 are complementary techniques yielding similar molecular weight on the native biopolymers. However, AF4 shows unique advantages in characterization of biopolymer aggregation, as it yields better separation between unimer and aggregates, and has no anomalous elution of aggregates. Also, AF4 yields more accurate characterization of aggregation level because there is no stationary phase that could interact with analytes, and there should be less shear force imparted on the aggregates during fractionation, as compared to SEC separation. The comparison of AF4 versus SEC in thermal stability studies of three biopolymers presented in this chapter suggests that AF4 yields more reliable data when the applied SEC condition is not able to elute all aggregates and/or has late elution of aggregates.

Introduction

Biopolymers or proteins in pharmaceutical applications need careful formulation to maintain the bioactivity of the drug. Aggregation is one of the more common degradation pathways of biopolymers, and in some cases, the aggregates can cause serious undesired biological response. Additives are used in the formulation to offer protection or stabilize the biopolymers against certain types of stress (*e.g.*, heat, cold, oxidation etc.). However, some of the additives can undergo auto-degradation. For example, polysorbates can undergo auto-oxidation, resulting in the formation of hydroperoxide (*1–3*), which in turn can oxidize a protein in the formulation. As such, monitoring molecular weight changes of biopolymer in the formulation is essential and can enable the development of novel additives to improve the stability of biopolymers in formulations.

Size-exclusion chromatography (SEC) and asymmetric flow field-flow fractionation (AF4) are widely employed techniques for characterization of biopolymers and their aggregation (*4–10*). However, both SEC and AF4 could have challenges in the separation and characterization of biopolymers and aggregates. Particularly in SEC, the stationary phase of SEC columns can interact with biopolymers and their aggregates. Also, SEC could have issues in the separation of ultrahigh molecular weight aggregates, weakly bound complexes, and highly branched polymer chains (*11*). On the other hand, AF4 could also have the interaction between membrane and analytes (*12*, *13*), or small polymer chains can be lost through the membrane if the polymer is too small. However, AF4 is a separation technique based on open-channel field flow fractionation that has much less shear force to break down biopolymer complexes and aggregates during fractionation. As such, AF4 would be well suited for ultrahigh molecular weight polymer chains. In this study, SEC and AF4 conditions were optimized for characterization of biopolymers, including bovine serum albumin (BSA), abatacept, and immunoglobulin G (IgG), in their native states and after heat stress in the presence and absence of thermal stabilizer. Detailed comparisons of SEC versus AF4 on the characterization of these biopolymers will be presented and discussed.

Experimental

Materials

Biopolymers: Abatacept is an IgG-derived drug purchased from a local pharmacy store (Great Lakes Pharmacy of Midland). It is a liquid solution having ~ 250 mg in 10 mL per vial or nominal concentration ~ 25 mg/mL; Immunoglobulin G (IgG, > 95%) and Bovine serum albumin (BSA, > 98% unimer) were purchased from Sigma-Aldrich. All biopolymers were used as received.

Biopolymer gel filtration standard (#1511901) was purchased from Bio-Rad Laboratories, Inc. (2000 Alfred Nobel Dr., Hercules, CA 94547), and used as received. The standard is a mixture of five biopolymers having reported molecular weights of 670, 158, 44, 17, 1.35 kDa, according to the product brochure.

Non-ionic surfactants M and F were proprietary materials synthesized by Dow Chemical and used as received.

All other reagents and buffers were of analytical or HPLC grade procured from Sigma-Aldrich or Fisher Scientific.

Sample Preparation

Biopolymer (BSA, abatacept and IgG) sample concentrations and injection volumes are given in Table 1. These sample solutions were prepared in the SEC eluent (pH 7 100 mM sodium phosphate buffer (NaPB) + 200 mM NaCl). The non-ionic surfactants having concentration ~ 0.1 mg/mL were prepared in the biopolymer solutions. Heat stress was carried out by heating solutions at various temperatures (40 to 70 °C) and time (5 minutes, 5 hours, or 22 hours). Solutions were centrifuged at 12,500 X *g* at ambient temperature for 1-5 minutes prior to SEC and AF4 injection.

Table 1. Sample Concentration and Injection Volumes

	AF4		*SEC*	
Biopolymers	*Concentration, mg/mL*	*Injection Volume, μL*	*Concentration, mg/mL*	*Injection Volume, μL*
BSA	0.4	50	0.4	20
Abatacept	0.4	20	0.4	10
IgG	2.0	20	2.0	10

SEC Conditions

An Agilent 1260 HPLC system equipped with a filter consisting of one layer of 0.1-µm nylon membrane, installed prior to the injection valve, was set at flow rate 0.4 mL/min. The injection was performed by an Agilent 1260 autosampler programmed to inject 10-20 µL of solution. SEC separation was carried out using one Waters BioSuite ultra high resolution (UHR) SEC column 4.6 mm i.d. x 300 mm length (nominal particle size 4 µm and pore size 250 Å) operated at 28 °C. SEC eluent was pH 7 100 mM sodium phosphate buffer (NaPB) + 200 mM NaCl. The detectors were an Agilent Diode Array detector (DAD) (280nm) connected with a Wyatt DAWN HELEOS II multi-angle laser light scattering (MALLS) detector and a Wyatt Optilab TrEX differential refractive index (DRI) detector. The Wyatt DAWN was equipped with a red laser operating at a wavelength of 660 nm and operated at ambient temperature, and TrEX DRI detector was equipped with a light emitting diode (LED) of the same wavelength and operated at 28 °C.

AF4 Conditions

AF4 fractionation shared the same Agilent 1260 LC pump, autosampler and detectors as those used in SEC. The injection volume was 20-50 µL of solution. AF4 was performed on an Agilent 1260 LC pump combined with a Wyatt DualTech Eclipse. For the separation, a short channel equipped with a 350 µm spacer of medium width and a regenerated cellulose membrane with a cutoff of 10 kDa was used. The mobile phase was pH 7 17 mM NaPB + 33 mM NaCl solution. The flow programs (cross and channel flow) of AF4 employed for individual biopolymer are given in Figure 1, and will be discussed in the Results and Discussion section.

Data Processing and Software

The Wyatt DAWN MALLS was calibrated with HPLC grade toluene filtered through a 0.05 µm syringe filter. The individual light scattering detectors were normalized to the 90° light scattering detector using bovine serum albumin (BSA) unimer as the isotropic scatter. The inter-detector delay volume was determined by aligning the 90° light scattering and DRI peaks obtained for BSA.

A dn/dc of 0.185 mL/g value was used for all biopolymers (BSA (*14*, *15*), abatacept (*16*) and IgG (*6*, *17*)). Signals from the MALLS and DRI detectors were analyzed using ASTRA 6 software, version 6.1. *Molar mass (M)* was determined at each SEC elution volume increment. To obtain these metrics, Debye plots covering the angular variation in scattering intensity from detector angles including from 25.9° to 132.2° were least-squares fit to a first-order polynomial according to Zimm formalism. The mass recovery was estimated by the ratio of calculated mass, using dn/dc 0.185 mL/g, to the injected mass. The amount of aggregates in solution were estimated by the ratio of the calculated aggregate mass using dn/dc 0.185 mL/g to the calculated mass of both unimer and aggregates.

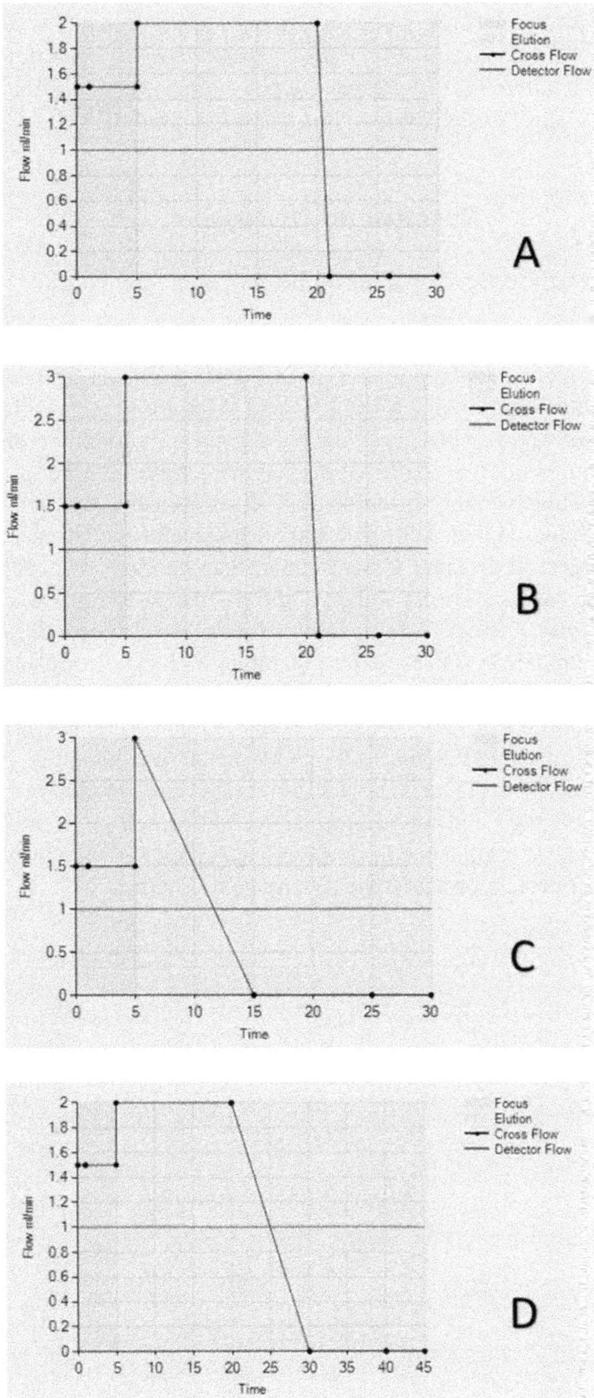
Focus
Elution
Cross Flow
Detector Flow
Flow ml/min
Time
A
Focus
Elution
Cross Flow
Detector Flow
Flow ml/min
Time
B
Focus
Elution
Cross Flow
Detector Flow
Flow ml/min
Time
C
Focus
Elution
Cross Flow
Detector Flow
Flow ml/min
Time
D

Figure 1. AF4 flow programs. Constant cross flow at flow rate 2.0 mL/min (A) for abatacept, and 3.0 mL/min (B) for BSA, Linear decay of cross flow from 3.0 to 0 mL/min in 10 min (C) for abatacept, and Combination of constant cross flow and linear decay of cross flow (D) for IgG.

Results and Discussion

SEC and AF4 Method Development on Biopolymer Standards

A biopolymer gel filtration standard was used for evaluation of the separation power of biopolymer (globular polymer) in SEC and various AF4 flow conditions. The UV chromatograms obtained by SEC and AF4 are given in Figure 2 and Figure 3, respectively. There are five biopolymers reported in the product specification from vendor, and all five biopolymers were observed in SEC. But the lowest molecular weight standard (1.35 kDa) eluted in the total inclusion volume of SEC column. In AF4, The lowest molecular weight standard (1.35 kDa) was not observed, because it was smaller than the pore size of the channel membrane which has a molecular weight cutoff of 10 kDa and hence, was filtered out in the separation channel. Additionally, various AF4 flow conditions yielded different separation power (or resolution) on either high or low molecular weight biopolymers. An AF4 condition having constant cross flow rate 3.0 mL/min yields better resolution in two lower molecular weight proteins, while a lower constant cross flow rate (2.0 mL/min) yields better resolution in two higher molecular weight proteins. The linear decay of cross flow program provides good resolution for both high and low molecular weight proteins. In addition, AF4 theory can be used as a guidance to develop a specific cross-flow program for a target biopolymer/protein based on the size range of interest.

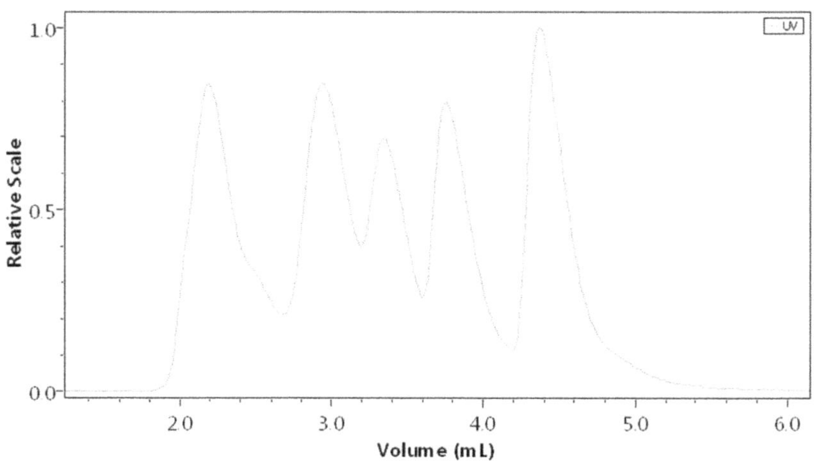

Figure 2. UV chromatograms of biopolymer gel filtration standard obtained by SEC.

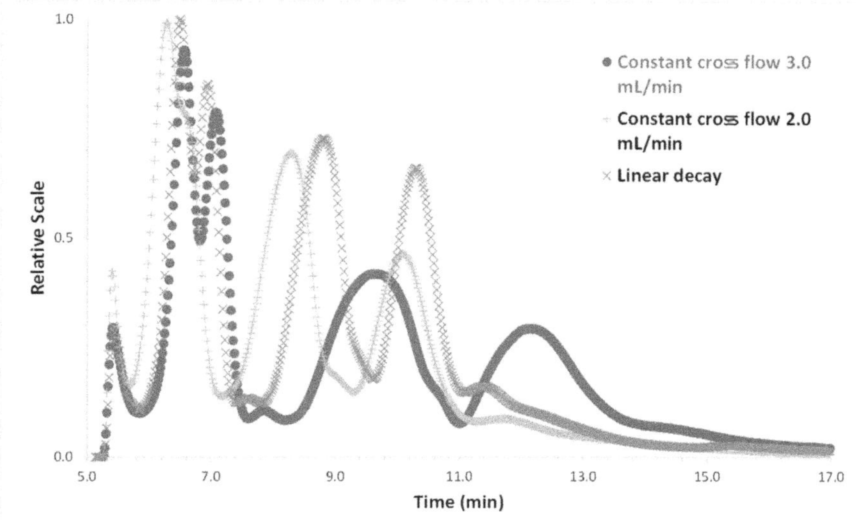

Figure 3. UV chromatograms of biopolymer gel filtration standard obtained by AF4 using constant cross flow rate 3.0 mL/min (red), constant cross flow rate 2.0 mL/min (green), and linear cross flow decay from 3.0 to 0 mL/min in 10 min (blue).

SEC and AF4 Method Development and Thermal Stabilization Study on BSA

BSA is a well-characterized protein and usually used as a protein standard to check instrument and calibration constants. As such, a BSA native solution (without heat stress) was characterized to ensure the instrument and separation were working properly. The obtained molecular weight along with LS (90°) chromatogram on a BSA solution by SEC and AF4 are given in Figure 4, and the obtained molecular weight averages are given in Table 2. A constant cross flow rate 3.0 mL/min was used in the AF4 separation of BSA. The obtained molecular weight averages, mass fraction of unimer, and mass recovery (Table 2) were almost identical by both SEC and AF4 methods. These results suggest that both AF4 and SEC conditions were able to separate dimers, trimers and oligomers from the BSA unimer. But, on closer comparison of the two chromatograms, AF4 appears to yield a better resolution in the separation of oligomers from unimer. The SEC column used in this study was Waters Biosuite ultra high resolution SEC column that is specifically for size separation of proteins and biopolymers. According to Waters column brochure, this column is able to provide size separation of a globular protein having molecular weight range of 10 to 500 kDa, which should be good for the separation of BSA unimer and oligomers. Further optimization of SEC conditions such as sample concentration and injection volume could improve the separation of BSA unimer and oligomers. A light scattering impurity peak around column exclusion volume (~ 1.5 mL) was observed. This impurity peak was not BSA aggregates, because eluent blank injections prior to BSA injections also showed this impurity peak with similar intensity. This impurity peak could interfere with the separation of higher molecular weight oligomers. As such, future study to investigate the source of this impurity peak and to eliminate it could further improve the separation of BSA.

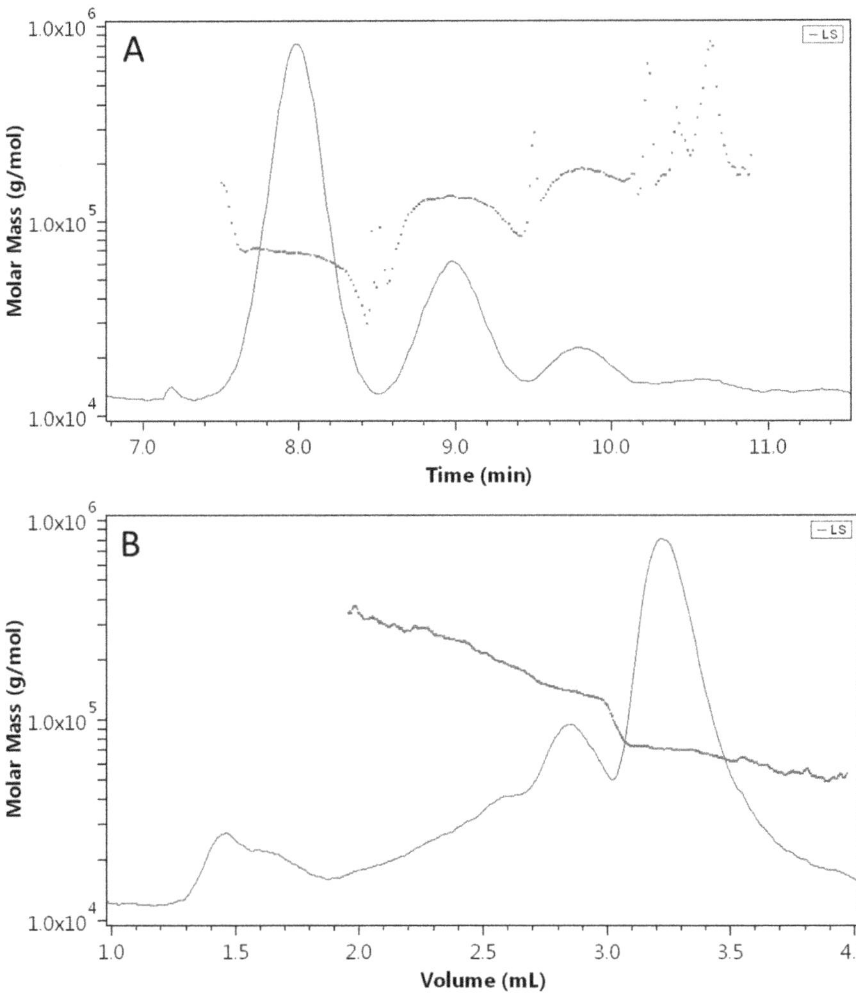

Figure 4. Molecular weight and LS (90°) chromatograms of BSA obtained by AF4 (A) and SEC (B).

Table 2. The Obtained Molecular Weight Averages (kDa) of BSA by SEC and AF4

	M_w	M_w/M_n	*Mass recovery (unimer+oligomers)*
BSA-by AF4	66.9	1.02	93.1%
BSA-by SEC	68.4	1.02	97.1%

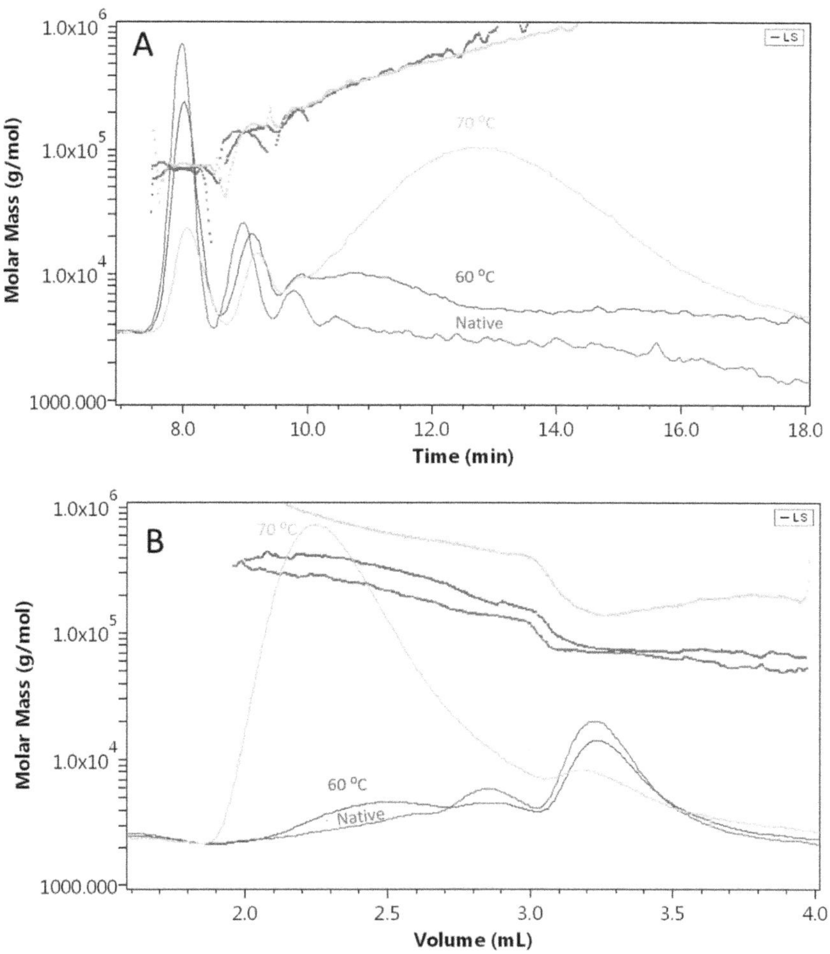

Figure 5. Molar mass along with LS (90°) chromatograms on BSA native (red), 60°C for 30 min (blue) and 70°C for 30 min (green) obtained by AF4 (A) and SEC (B).

Table 3. The Obtained M_w and Mass Fraction on BSA Unimer and Oligomers by AF4

	Unimer		*Dimer*		*Higher Aggregates*		
	M_w (kDa)	*Mass fraction (%)*	*M_w (kDa)*	*Mass fraction (%)*	*M_w (kDa)*	*Mass fraction (%)*	*Mass recovery (%)*
BSA Native	66.9	75.4	121	17.3	214	7.3	93.1
BSA-Heated at 60 °C	69.3	66.0	135	16.3	369	17.7	93.9
BSA-Heated at 70 °C	73.6	32.2	151	12.2	608	55.6	89.6

Table 4. The Obtained M_w (kDa) and Mass Fraction on BSA Unimer and Oligomer by SEC

	Unimer		*Dimer and Aggregates*		
	M_w	*Mass fraction (%)*	M_w	*Mass fraction (%)*	*Mass recovery (%)*
BSA Native	68.4	79.6	174	20.8	97.1
BSA – Heated at 60 °C	76.6	80.2	259	19.8	78.9
BSA – Heated at 70 °C	181	49.4	714	51.4	57.1

Table 5. The Obtained M_w (kDa) and Mass Fraction on BSA Solutions That Were Heated at 60 °C for 30 Minutes with or without M by AF4

	Unimer		*Dimer*		*Aggregates*		
	M_w	*Mass fraction (%)*	M_w	*Mass fraction (%)*	M_w	*Mass fraction (%)*	*Mass recovery (%)*
BSA without M	69.3	66	135	16.3	369	17.7	93.9
BSA with M	66.6	72	128	19.5	435	8.5	94.6

The obtained molar mass along with LS (90°) chromatograms on heat-stressed BSA solutions by AF4 and SEC are given in Figure 5. The obtained M_w, mass fraction and mass recovery are given in Tables 3 and 4, respectively. These results again suggest that AF4 has better resolution in the separation of unimer from dimer and oligomer. The obtained M_w of the BSA unimer had almost no changes for the samples having ultrahigh molecular weight aggregates (heat-stressed samples). But the obtained SEC M_w of the unimer of the heat-stressed solutions were elevated due to the late elution of ultrahigh molecular weight aggregates. The late elution of ultrahigh molecular weight chains were only occurred in a non-ideal SEC, but not in AF4, because the low molecular weight unimers elute earlier than ultrahigh molecular weight aggregates by field-flow fractionation. The late elution of ultrahigh molecular weight materials in SEC separation was reported in other polymer systems, including branched polymers, and ultrahigh molecular weight materials (*11*, *18–20*). The exact mechanism of late elution cannot be completely discerned from the data shown in Figures 5. Several competing mechanisms have been discussed by Giddings (*21*), but virtually all of them lead to late elution of ultrahigh molecular weight chains, and usually it is very challenging to overcome this non-ideality of SEC. As such, no further optimization of SEC condition was attempted, since it was out of the scope of this study. Additionally, lower mass recoveries were also observed on the heat-stressed solutions (60 and 70 °C) by SEC, suggesting ultrahigh molecular weight aggregates in heat-stressed solutions did not elute from the SEC column. Therefore, AF4 is a much better technique in characterization of BSA solutions having ultrahigh molecular weight aggregates.

The enhancement of thermal stability of BSA solutions by non-ionic surfactant M was also studied. BSA sample solutions were heated at 60 °C for 30 minutes in presence or absence of M. The results as given in Figure 6 and Table 5 suggest that M is able to stabilize BSA during heat stress of the solutions. Much lower aggregate content was observed in the presence of non-ionic surfactant M. M_w of aggregates (Table 5) should be used cautiously, because it could have large errors due to low S/N of RI signal.

SEC and AF4 Method Development and Thermal Stabilization Study of Abatacept

AF4 cross flow conditions were studied to optimize the separation of abatacept. The results (Figure 7) suggest that AF4 using linear decay of cross flow yields a better separation (narrow peak with higher S/N), since abatacept is a mono-dispersed protein and should be a narrow peak. As such, it was decided to use linear decay of cross flow condition for further study of abatacept.

AF4 and SEC techniques were employed for the characterization of native and thermally aged abatacept solutions that were held at various temperatures for 22 hours. Molecular weight and LS (90°) chromatograms are presented in Figure 8, and the obtained M_w of these solutions are given in Table 6. These results suggest that AF4 and SEC yield similar results for the characterization of native abatacept. Note the mass recoveries were higher than 100%, which could be due to inaccuracy in prepared concentrations, because abatacept sample solutions were directly diluted from a purchased stock solution having nominal concentration ~ 25 mg/mL. As such, the recovery data on all abatacept studies should be used for relative comparison only.

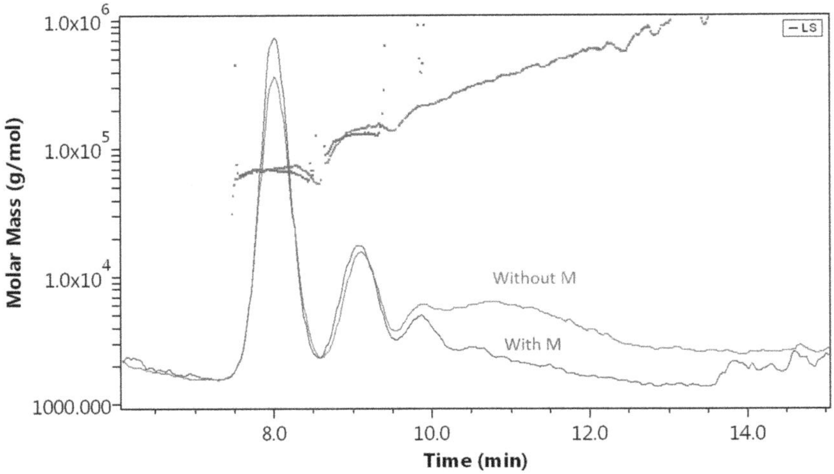

Figure 6. Molar mass along with LS (90°) of BSA solutions that were heated at 60 °C for 30 minutes with (blue) or wihtou (red) M obtained by AF4.

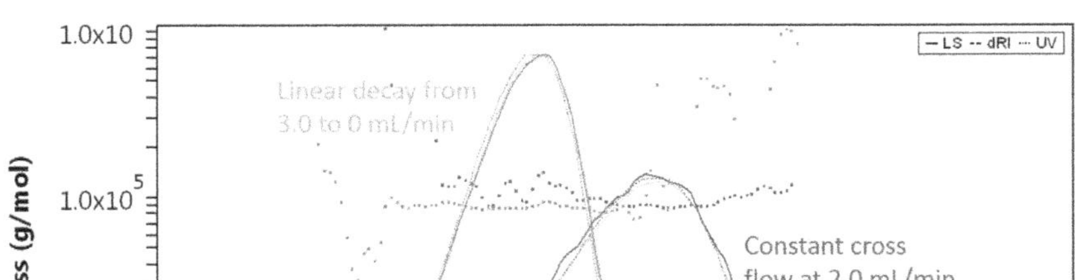

Figure 7. Molar mass and LS (90°) (solid), RI (dot) and UV (dash) obtained by AF4 using linear decay of cross flow from 3.0 to 0 mL/min (green) or constant cross flow at flow rate 2.0 mL/min (blue).

Also, these results indicate that abatacept was thermally degraded when it was heated for 22 hours. The thermal degradation was manifested by the appearance of high molecular weight aggregates. AF4 yields a higher resolution in the separation of high molecular weight aggregates from unimer, and the obtained M_w for the unimer was not affected by the aggregates. AF4 has no anomalous elution of ultrahigh molecular weight materials to interfere with the unimer peak. Such is not the case in SEC where aggregates elute anomalously late and thus interfere with the unimer peak. AF4 therefore offers a clear advantage for characterization of thermally degraded proteins.

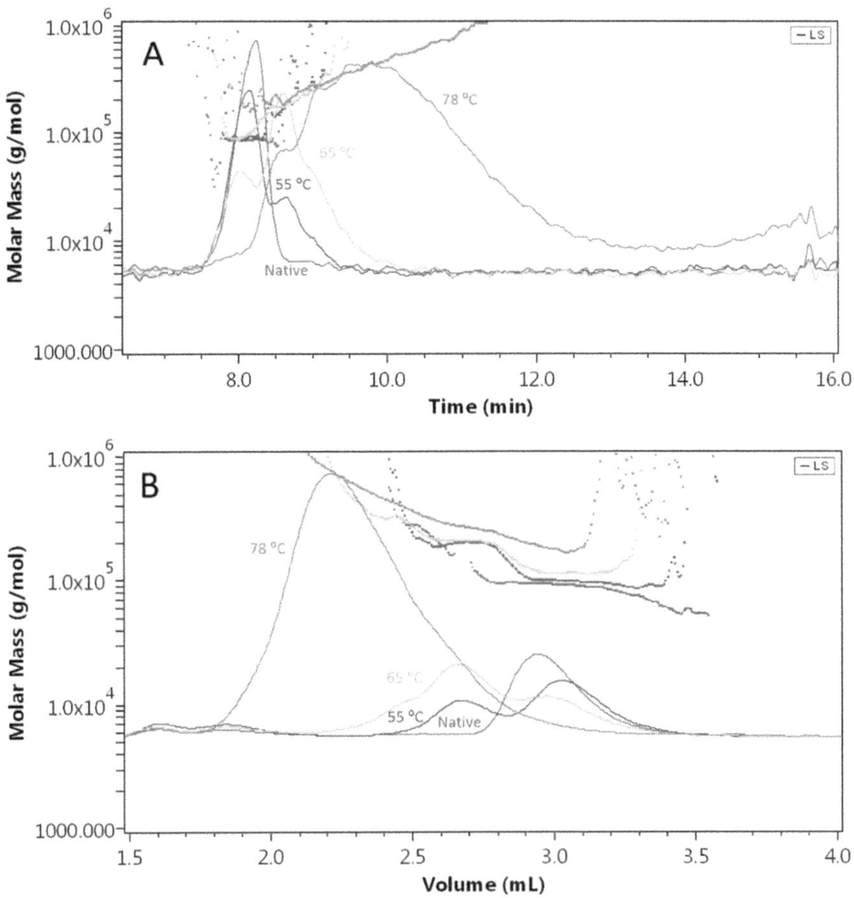

Figure 8. Molar mass along with LC (90°) obtained by AF4 (A) and SEC (B) for the separation of abatacept native (red), heated for 22 hours at 52 °C (blue), 65 °C (green) and 78 °C (magenta).

Table 6. The Obtained M_w (kDa) and Aggregate Fraction of Heated Abatacept Samples by AF4 and SEC

	By AF4				*By SEC*			
	Unimer M_w	*Aggregates M_w*	*Aggregates (%)*	*Mass Recovery (%)*	*Unimer M_w*	*Aggregates M_w*	*Aggregates (%)*	*Mass Recovery (%)*
Abatacept, native	87.1	NA	NA	110	90.3	NA	NA	131
Abatacept 52 °C	96.3	192	17	110	101	214	18.8	113
Abatacept 65 °C	99.6	236	57	107	125	236	48.3	108
Abatacept 78 °C	NA	489	100	128	NA	568	100	132

The thermal stability of abatacept in the presence of non-ionic surfactant F was investigated and the results are given in Figure 9 and Table 7, respectively. The results suggest that SEC and AF4 are in good agreement yielding similar M_w values and aggregate contents. Also, these results indicated that F was able to stabilize the abatacept solution for thermal stress The heat-stressed abatacept solution containing stabilizer had much lower aggregate content than the solution without stabilizer, suggesting the solution in the presence stabilizer F has much less thermal degradation.

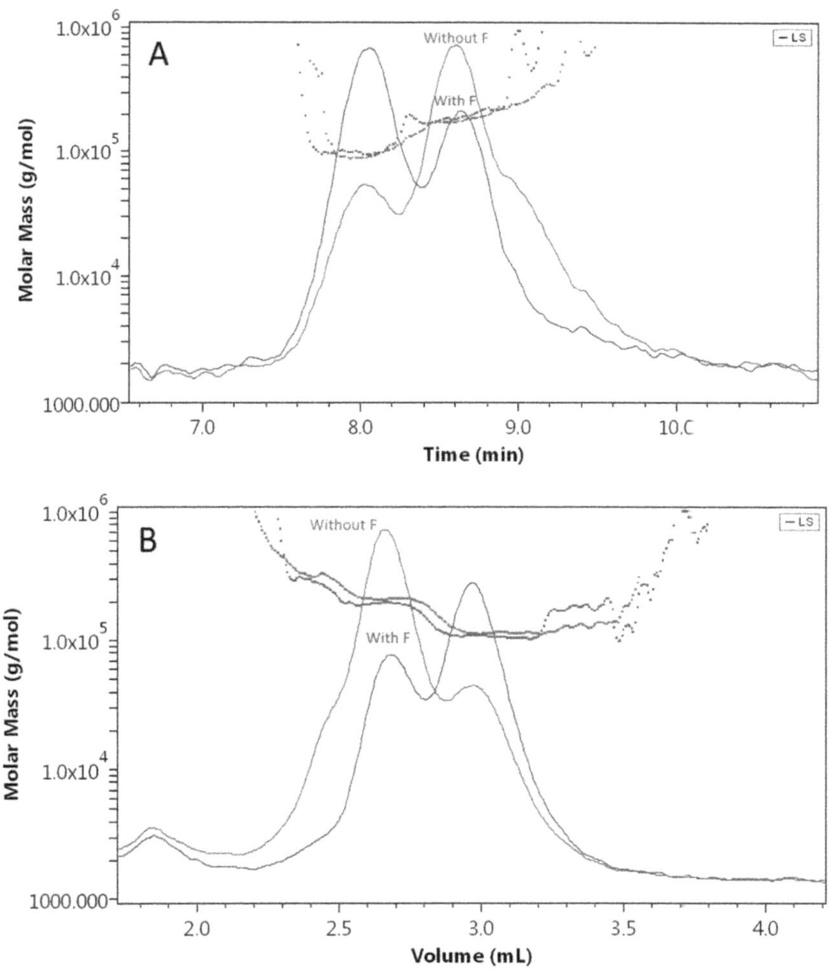

Figure 9. Molar mass along with LC (90°) obtained by AF4 (A, and SEC (B) on abatacept solutions that were heated at 65°C for 22 hours with (blue) or without F (red).

Table 7. M_w (kDa) and Aggregate Fraction Obtained by AF4 versus SEC on Abatacept Solutions That Were Heated at 65°C for 22 Hours in the Presence and Absence of F

	By AF4				*By SEC*			
	Unimer M_w	*Aggregates M_w*	*Aggregates (%)*	*Mass Recovery (%)*	*Unimer M_w*	*Aggregates M_w*	*Aggregates (%)*	*Mass Recovery (%)*
Abatacept	96.3	236	57	106	125	236	48	108
Abatacept with F	118	265	31	118	128	210	23	124

SEC and AF4 Method Development and Thermal Stabilization Study of IgG

The AF4 cross flow program for the characterization of IgG solutions was selected according to a previous publication (*6*). An injection volume study on an IgG solution at a concentration ~ 2 mg/mL was performed, and the results are given in Figure 10. The injection volume of 20 µL was selected for further studies, as this condition yields the best tradeoff between S/N ratio and resolution.

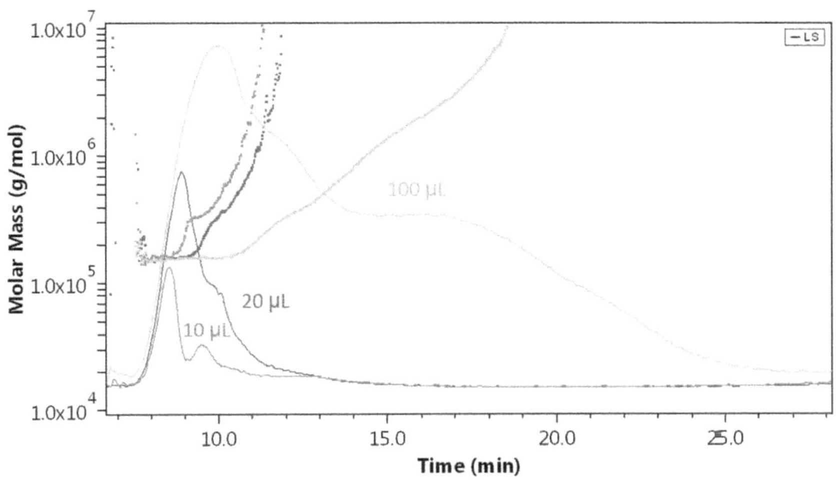

Figure 10. Injection volume study on an IgG solution (~ 2 mg/mL). Molar mass along with LS (90°) on the injection volumes of 10 (magenta), 20 (blue) and 100 µL (green) by AF4.

In this study, two batches of IgG samples named IgGBB and IgGE2 were used. These two batches were purchased at different times and had different storage times in a refrigerator (4 °C). The amounts of aggregates observed in these two batches were different (Figure 11). More aggregates were observed in the batch IgGBB. Both batches were used in the following studies based on the availability of the prepared IgG sample solutions.

Comparison of AF4 and SEC characterization of native and heat-stressed IgG (60 °C for 30 minutes) is given in Figure 12 and Table 8. The results indicate both SEC and AF4 are good techniques and yield very similar results for M_w and mass fraction of aggregates. The mass fraction of unimer and aggregates were only estimated, since there was no baseline separation between unimer and aggregates by either SEC or AF4 condition.

The mass recovery by both SEC and AF4 were similar, but were about 80%. The purchased IgG samples may contain impurities and/or buffers, so the prepared concentration may not be accurate. However, recovery data are still valid for relative comparison.

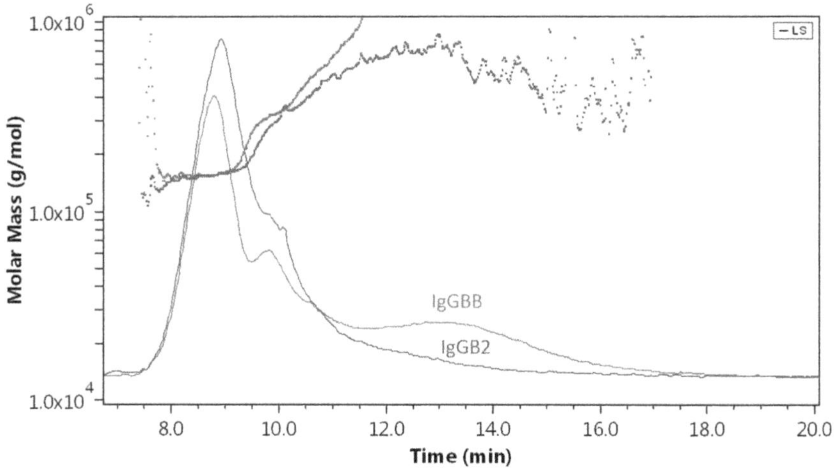

Figure 11. Molar mass along with LS (90°) on IgG (batch BB) (red) and IgG (batch B2) (blue) obtained by AF4.

Table 8. The Obtained M_w (kDa) and Mass Fractions on IgG Unimer and Aggregates by SEC and AF4

	Unimer		*Aggregates*		
	Mw	*Mass fraction (%)*	*Mw*	*Mass fraction (%)*	*Mass recovery*
SEC IgG, Native	154	84	444	16	76%
SEC IgG, Heated at 60 °C	172	72	1952	28	80%
AF4 IgG, Native	160	82	404	18	78%
AF4 IgG, Heated at 60 °C	156	79	1225	21	69%

Effect of non-ionic surfactant F on the thermal stabilization of IgG was investigated. As aforementioned, the F was able to provide thermal stability enhancement for abatacept solutions. However, the results shown in Figure 13 and Table 9 suggest that the addition of non-ionic surfactant F did not provide thermal stability enhancement for IgG solutions. This result could indicate the amount of addition of F was not enough to provide thermal stability enhancement for IgG, and further concentration study of F on thermal stability of IgG is needed.

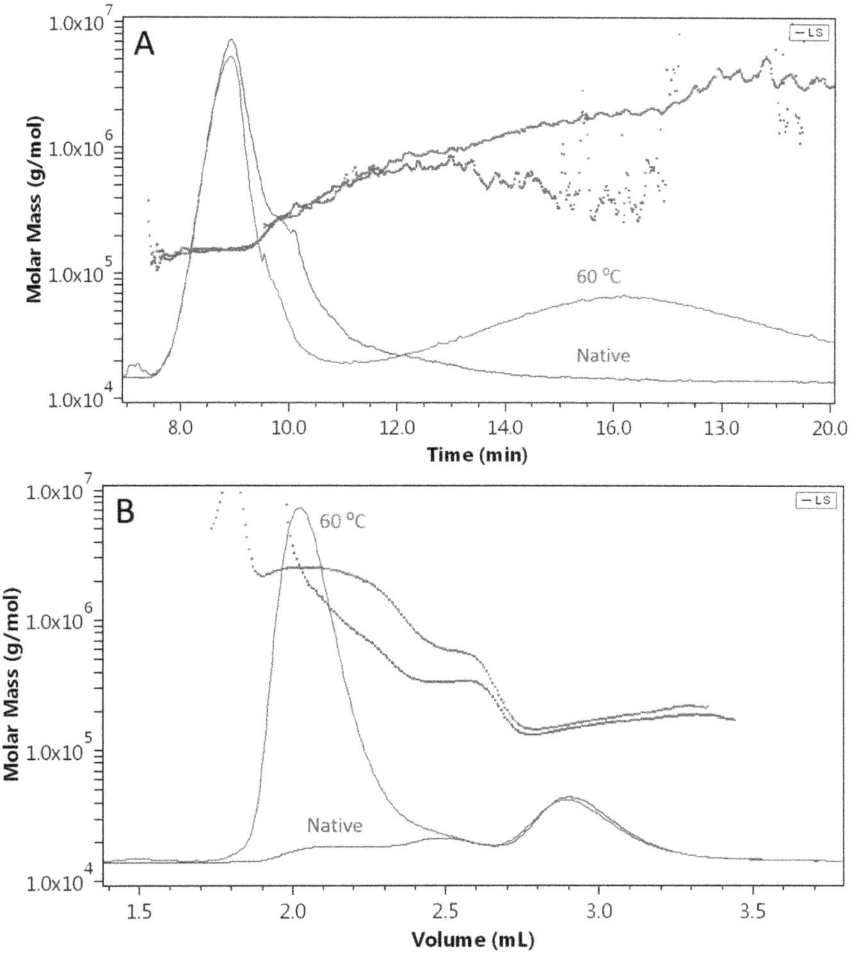

Figure 12. Molar mass along with LS (90°) chromatograms on heat-stressed IgG (red, 60°C for 30 minutes) and native IgG (blue) samples obtained by AF4 (A) and SEC (B).

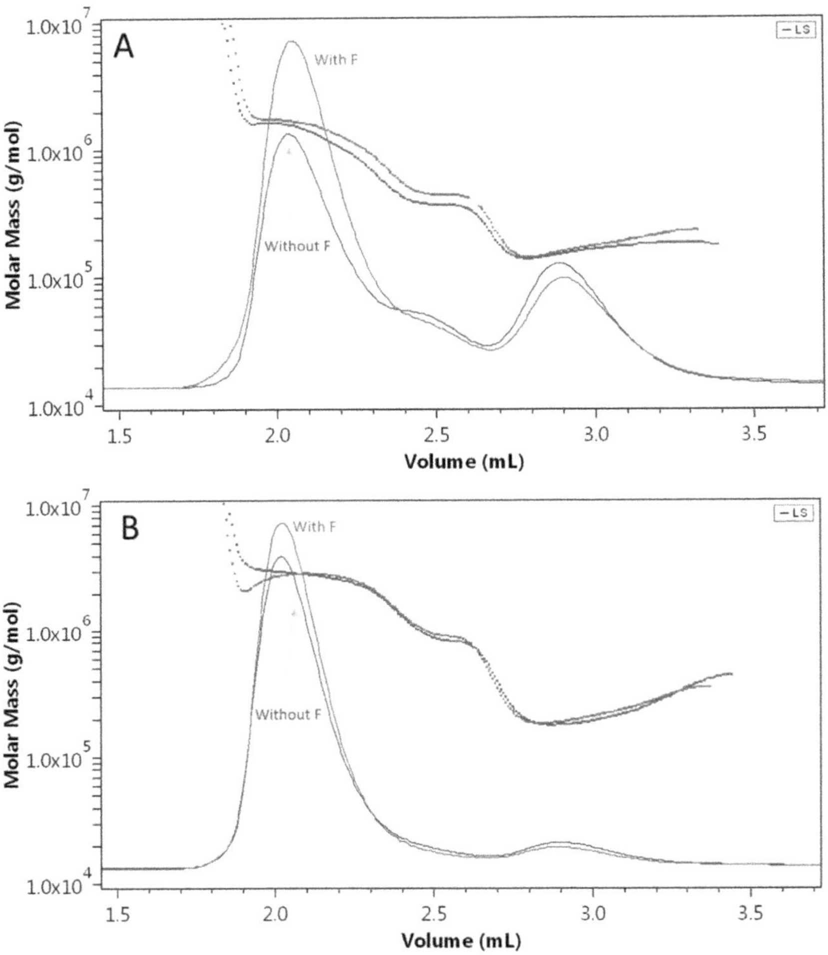

Figure 13. SEC characterization of heat-stressed IgG solutions with (red) and without (blue) F. Molar mass along with LS (90°) on IgG solutions that were heated at 40°C (A) and 60°C (B) for 30 minutes obtained by SEC.

Table 9. The Obtained M_w (kDa) on Heat-Stressed IgG Heated at 40 °C and 60 °C with and without F by SEC

	Unimer		*Aggregates*		*Mass recovery*
	Mw	*Mass fraction (%)*	*Mw*	*Mass fraction (%)*	
IgG, heated at 40 °C	164	74	894	26	83%
IgG, heated at 40 °C with F	175	70	1171	30	71%
IgG, heated at 60 °C	231	60	2581	40	84%
IgG, heat at 60°C with F	228	50	2509	50	75%

Conclusions

SEC and AF4 conditions were developed for characterization of three biopolymers and their aggregates for thermal stability study. The obtained results suggest that SEC and AF4 are complementary techniques yielding similar molecular weights on the native state of biopolymers. However, AF4 shows unique advantages in characterization of biopolymer aggregation, as it yields better separation between unimer and aggregates, and has no anomalous elution of aggregates. Also, AF4 is anticipated to yield more accurate characterization of aggregation level because the shear force is not expected to break down the aggregates during fractionation, and it has no stationary phase to interact with aggregates, as compared to SEC separation. These studies demonstrate that AF4 and SEC are useful analytical tools for the characterization of biopolymers. AF4 appears to be better suited for characterization of thermally degraded proteins, as it provides better resolution than SEC and does not suffer from anomalous elution of aggregates like SEC. AF4 and/or SEC yield essential information for development of various additives to improve stability of biopolymers in formulations.

References

1. Kerwin, B. A. Polysorbates 20 and 80 used in the formulation of protein biotherapeutics: structure and degradation pathways. *J. Pharm. Sci.* **2008**, *97*, 2924–2935.
2. Lee, H. J.; McAuley, A.; Schilke, K. F.; McGuire, J. Molecular origins of surfactant-mediated stabilization of protein drugs. *Adv. Drug Delivery Rev.* **2011**, *63*, 1160–1171.
3. Kishore, R. S. K.; Pappenberger, A.; Dauphin, I. B.; Ross, A.; Buergi, B.; Staempfli, A.; Mahler, H.-C. Degradation of polysorbates 20 and 80: Studies on thermal autoxidation and hydrolysis. *J. Pharm. Sci.* **2011**, *100*, 721–731.

4. Fekete, S.; Beck, A.; Veuthey, J.-L.; Guillarme, D. Theory and practice of size exclusion chromatography for the analysis of protein aggregates. *J. Pharm. Biomed. Anal.* **2014**, *101*, 161–173.
5. Goyon, A.; Beck, A.; Colas, O.; Sandra, K.; Guillarme, D.; Fekete, S. Evaluation of size exclusion chromatography columns packed with sub-3μm particles for the analysis of biopharmaceutical proteins. *J. Chromatogr. A* **2017**, *1498*, 80–89.
6. Ma, D.; Martin, N.; Tribet, C.; Winnik, F. M. Quantitative characterization by asymmetrical flow field-flow fractionation of IgG thermal aggregation with and without polymer protective agents. *Anal. Bioanal. Chem.* **2014**, *406*, 7539–7547.
7. Bria, C.; Violleau, F.; Kim, S.; Williams, R. Field-flow fractionation for biological, natural, and synthetic polymers: recent advances and trends. *LCGC North Am.* **2013**, *31*, 1008, 1010–1017.
8. Yohannes, G.; Jussila, M.; Hartonen, K.; Riekkola, M. L. Asymmetrical flow field-flow fractionation technique for separation and characterization of biopolymers and bioparticles. *J. Chromatogr. A* **2011**, *1218*, 4104–4116.
9. Huerzeler, C.; Klein, T. In *Asymmetric FFF in biotechnology*; CRC Press: 2010; pp 136-138.
10. Williams, S. K. R.; Lee, D. Field-flow fractionation of proteins, polysaccharides, synthetic polymers, and supramolecular assemblies. *J. Sep. Sci.* **2006**, *29*, 1720–1732.
11. Podzimek, S.; Vlcek, T.; Johann, C. Characterization of branched polymers by size exclusion chromatography coupled with multiangle light scattering detector. I. Size exclusion chromatography elution behavior of branched polymers. *J. Appl. Polym. Sci.* **2001**, *81*, 1588–1594.
12. Jochem, A.-R.; Ankah, G. N.; Gerstner, D.; Elsenberg, S.; Roesch, U.; Johann, C.; Kraus, T. In *Colloidal interactions during asymmetric flow field-flow fractionation: Particle loss mechanisms studied with gold nanoparticles*; American Chemical Society, 2015; pp COLL-444.
13. Wagner, M.; Holzschuh, S.; Traeger, A.; Fahr, A.; Schubert, U. S. Asymmetric Flow Field-Flow Fractionation in the Field of Nanomedicine. *Anal. Chem.* **2014**, *86*, 5201–5210.
14. *Polymer Handbook*, 3rd ed.; John Wiley & Sons: New York, 1989; pp VII/641.
15. Yohannes, G.; Wiedmer, S. K.; Elomaa, M.; Jussila, M.; Aseyev, V.; Riekkola, M.-L. Thermal aggregation of bovine serum albumin studied by asymmetrical flow field-flow fractionation. *Anal. Chim. Acta* **2010**, *675*, 191–198.
16. Fast, J. L.; Cordes, A. A.; Carpenter, J. F.; Randolph, T. W. Physical Instability of a Therapeutic Fc Fusion Protein: Domain Contributions to Conformational and Colloidal Stability. *Biochemistry* **2009**, *48*, 11724–11736.
17. Hawe, A.; Romeijn, S.; Filipe, V.; Jiskoot, W. Asymmetrical flow field-flow fractionation method for the analysis of submicron protein aggregates. *J. Pharm. Sci.* **2012**, *101*, 4129–4139.

18. Gerle, M.; Fischer, K.; Roos, S.; Mueller, A. H. E.; Schmidt, M.; Sheiko, S. S.; Prokhorova, S.; Moeller, M. Main Chain Conformation and Anomalous Elution Behavior of Cylindrical Brushes As Revealed by GPC/MALLS, Light Scattering, and SFM. *Macromolecules* **1999**, *32*, 2629–2637.
19. Porsch, B.; Welinder, A.; Korner, A.; Wittgren, B. Distribution analysis of ultra-high molecular mass poly(ethylene oxide) containing silica particles by size-exclusion chromatography with dual light-scattering and refractometric detection. *J. Chromatogr. A* **2005**, *1068*, 249–60.
20. Li, Y.; Shen, H.; Lyons, J. W.; Sammler, R. L.; Brackhagen, M.; Meunier, D. M. Size-exclusion chromatography of ultrahigh molecular weight methylcellulose ethers and hydroxypropyl methylcellulose ethers for reliable molecular weight distribution characterization. *Carbohydr. Polym.* **2016**, *138*, 290–300.
21. Giddings, J. C. Analysis of fundamental obstacles to the size exclusion chromatography of polymers of ultrahigh molecular weight. *Adv. Chromatogr.* **1982**, *20*, 217–58.

Chapter 9

Advanced AF4 Characterization of Dendritic Biomacromolecules, Their Self-Assembly, and Hybrid Formation

Albena Lederer,[*,1,2] Susanne Boye,[1] and Dietmar Appelhans[1]

**[1]Leibniz-Institut für Polymerforschung Dresden e.V.,
Hohe Strasse 6, 01169 Dresden, Germany
[2]Technische Universität Dresden, School of Science,
01062 Dresden, Germany
[*]E-mail: lederer@ipfdd.de.**

Functional dendritic polymers with emphasis on bioactive interactions increasingly draw the interest of the macromolecular community. These biosystems are characterized by highly complex composition, including self-assembled polymers, the ability to complex drugs and responsiveness against external stimuli. This complexity has to be considered when developing effective biomaterial. Yet, for most biopolymer systems standard analytical methods provide non-quantitative parameters, without information on the distribution of the components, which is essential for physiological applications. The need for detailed analysis of bioactive macromolecules led to the development of novel physicochemical approaches with high application potential. Asymmetric flow field flow fractionation (AF4) is considered a smooth separation technique with a variety of hyphenation possibilities suitable for this purpose. AF4 in combination with complementary techniques such as microscopy and simulations can additionally give essential quantitative details on the systems composition and its responsiveness.

Macromolecular architectures can be composed of monomeric units which differ in their chemical structure, number and coupling sequence. This results in a number of different possible macromolecular structures, which can exist in one sample at the same time featuring multiple structural distributions in molar mass, chemical structure, functionality type and branching topology. There are several approaches for the analysis of macromolecular distributions, which were developed in recent years. Size exclusion chromatography (SEC) coupled with different absolute or relative detection techniques is the typically used method for the characterization of molar mass. Two-dimensional liquid chromatography, in which SEC is coupled to liquid adsorption chromatography can deliver simultaneous information on molar mass and chemical distributions (*1*). Additionally, other approaches showed the potential of liquid chromatography and other fractionation techniques for separation according to branching (*2–7*). Understanding molecular structures and their distributions is essential for tuning the properties of functional soft matter. This issue is more complex when using multifunctional, highly branched structures. In addition to their structural characteristics, intelligent macromolecular materials can respond to external stimuli to fulfill their role as active systems either by self-assembly or by controlled interaction with other (macro)molecules or particles. These properties are essential in the development of bioactive materials as it was demonstrated in a number of publications (*8–17*). In our recent works we show the development of aggregation of giant glycopolymers (*18*), biohybrid formation between synthetic macromolecules and proteins (*19–22*), and drug complexation and release by glycodendritic hosts (*23*, *24*) (Figure 1). These systems are characterized by very complex composition, in regard to size and amount of different components: along with the single components, complexes, hybrid structures and self-assemblies with different structure parameters are formed, which additionally respond to e.g. pH conditions. The understanding of the formation mechanism, kinetics, as well as quantification of these structures is essential for their use in biorelated applications.

For the complete analysis of these properties we need a method which is gentle to weak, non-covalent interactions. Additionally, quantitative information on the different components in complex systems is needed. Furthermore, broad size distributions should be simultaneously detectable in order to achieve reliable information in a broad size region: from the molecular level (few nm) to that of particle/hybrid formation (few µm). Fast switching of aqueous conditions such as pH or ionic strength should be possible in order to guarantee detection of structural changes *in situ*. And finally, complete information on size, molar mass and conformation distribution under changing conditions should be reliably extracted.

All these requirements can be potentially fulfilled by using asymmetric flow field flow fractionation (AF4) (*25–27*) as an alternative separation technique to SEC. AF4 separation takes place in a channel without the need of column material. Therefore, interactions and strong shear forces during the separation process are reduced compared to column based separation. Broad size distributions characterization is the main advantage of AF4 with some limits in the low molecular weight region because of the limited cut-off of the membrane in the channel (Figure 2).

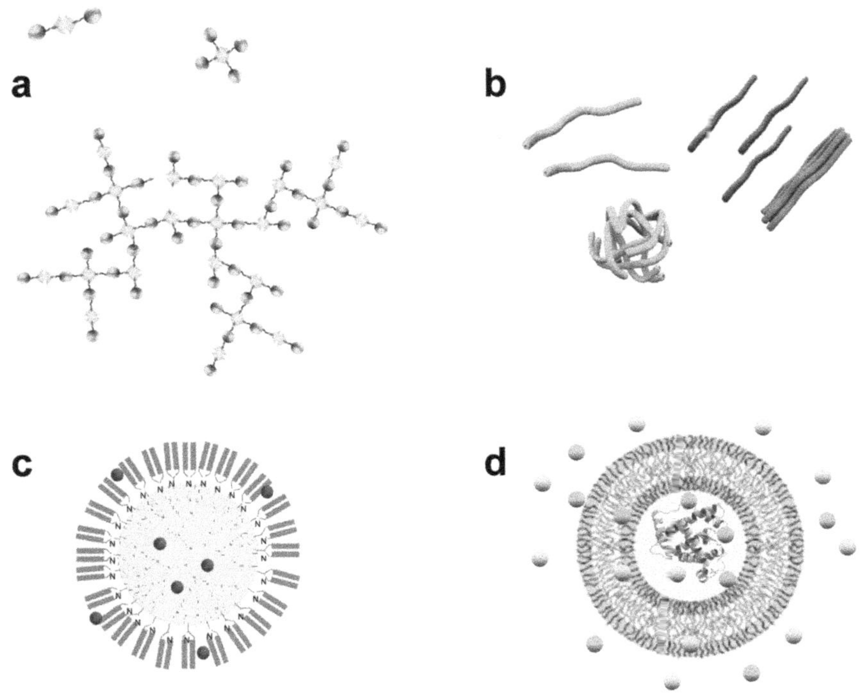

Figure 1. Examples of different interactions of synthetic macromolecules and in biologically relevant systems as:biohybrid structures with proteins (a), supramolecular structures of glycomacromolecules (b), drug carrier systems of glycodendrimers(c), cell-mimicking polymersomes (d).

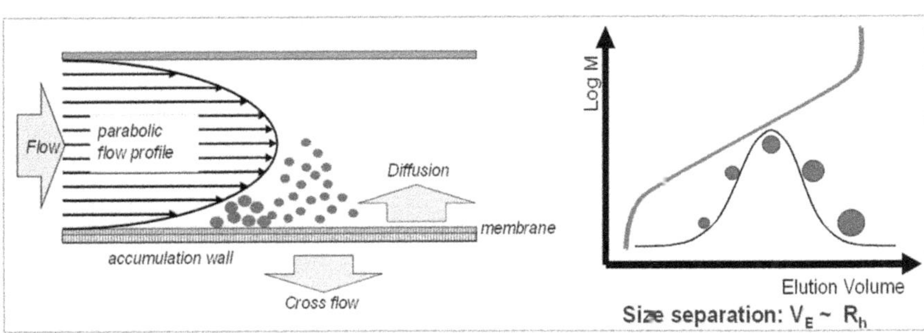

Figure 2. Separation principle of AF4: Scheme of channel profile with horizontal channel flow and perpendicular forces during the separation of a sample mixture, and the resulting separation according to hydrodynamic properties.

Yet, separation alone does not deliver size and conformational distributions. For this purpose an adequate detection after separation is needed. Static light scattering detection, for example, is needed to obtain information on absolute molar mass and radius of gyration. Furthermore, dynamic light scattering can deliver the hydrodynamic radius. Another possibility to obtain this value is the application of special interpretation of the separation process (*28*), which is based on the diffusion characteristics of differently sized particles, as shown by Nilsson et al. (*29–31*) UV-Vis and differential refractive index detectors can be useful to achieve concentration information either for further interpretation of the light scattering data or for selective quantification of sensitive species.

The information on radius of gyration and hydrodynamic radius together with the molar mass leads to powerful analysis of the conformation of complex systems after separation. Based on the scaling law

$$R_g = K \cdot M^{\nu} \qquad \text{(Equation 1)}$$

the scaling exponent ν can be estimated, which gives a first impression about the solution state of the single components and their association as the values of ν are well understood for polymers, or different geometrical objects, for example, ν = 0.33 corresponds to a hard sphere, n = 0.59 to a statistical coil and ν = 1 to a rigid rod (*32*).

Now, the combination of physical dimension and molar mass can additionally deliver density information and using this fact and equation 2, the apparent global density of the macromolecules and particulates can be estimated:

$$\rho_{app} = \frac{M_w}{V(R)} \qquad \text{(Equation 2)}$$

In contrast to the global density, which is an average over the whole macromolecule, the segmental density corresponds to the segmental distribution within the macromolecule. Additional information about the segmental density in the different components can be obtained from the combination of R_g and R_h. This is based on the fact that the radius of gyration is simply the average of the distances between the center of mass to any segment of the particle, while the hydrodynamic radius depends on the diffusion properties of the particle and corresponds to the diffusion of a sphere. Using these differences, a ratio of both radii delivers the ρ-parameter (equation 3), which is characteristic for different segmental densities, calculated and measured for sphere (0.775), rod (between 2.1 and 2.8), branched structures (1.23-1.73) or gels (0.3-0. 6) (*32*).

$$\rho = R_g/R_h \qquad \text{(Equation 3)}$$

The separation approach consisting of AF4 combined with light scattering was used in a number of investigations for understanding biomolecular systems and complex macromolecular architectures (*18–24*, *33–37*). An interesting class of biomacromolecules are the glycodendritic structures which show very high potential for applications in medicine, diagnostics, and biomimetics (*38–49*). The dendritic structures alone possess quite complex molecular architecture as shown in Figure 3 .

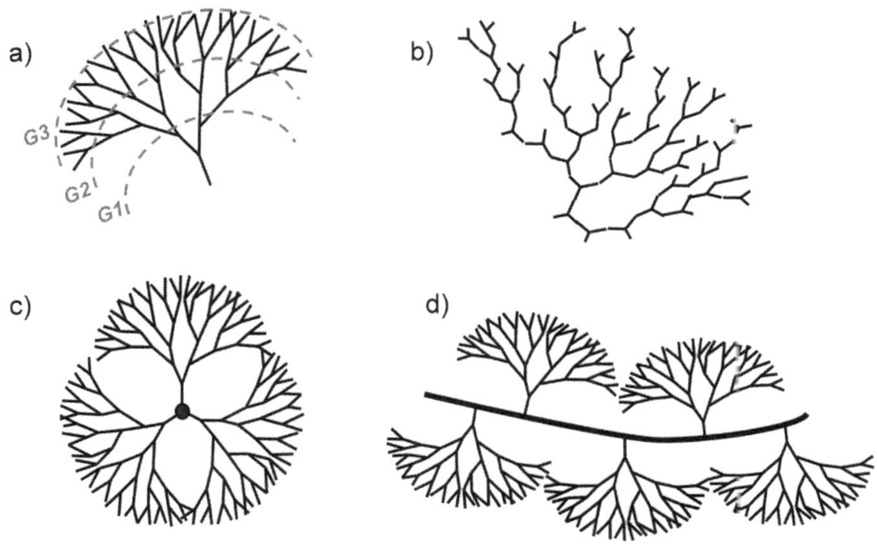

Figure 3. Schematic structures of dendritic macromolecular architectures: a) dendron in third generation, b) hyperbranched polymer, c) dendrimer and d) dendronized polymer.

The name dendritic comes from the type of branching, which resembles a tree (δέντρο, Greek). The branching is potentially possible at every monomer unit, which results in highly dense structures (*50*). Controlled stepwise synthesis leads to perfectly branched dendrons. When coupled to a core molecule a dendrimer with a highly defined, monodispersed structure is formed. When these dendrons are coupled to a linear macromolecular backbone, dendronised polymers (DenPols) are obtained. If the dendritic polymers are synthesized not by stepwise synthesis, but by a statistical reaction in one pot, the control over the dendritic structure is driven by a statistical process, leading to an imperfect branching in the so called hyperbranched polymers. As a result of their very dense branching, the dendritic polymers possess a compact structure with high macromolecular segment density compared to their linear analogs. Additionally, they are characterized by a very high amount of chain-ends, which can be purposefully modified for use in special applications. The modification of these end-groups with sugar units leads to glycodendritic structures, which improves solubility in aqueous media, reduces toxicity and enables interaction with proteins and other biologically relevant systems (*51*, *52*).

One prominent example of glycodendritic structures are DenPols formed by the coupling of lysine dendrons along a linear maleimide copolymer chain (*18*). Figure 4 depicts the structure of the linear chain (MI-G0) as well as the DenPols with different generation of the dendrons (MI-G1 to MI-G3).

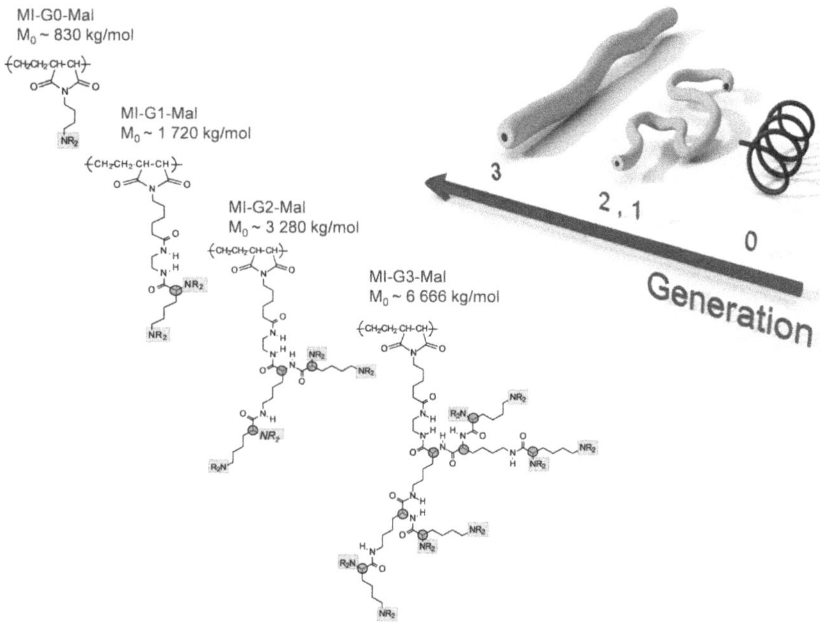

Figure 4. Chemical structure of dendronized polymers based on a maleimide copolymer backbone and lysine dendrons, which are additionally modified with maltose (R = maltose) in four different generations forming various molecular conformations. (reprinted with permission from Boye et al. Biomacromolecules 2012, 13 (12), 4222-4235, copyright (2012) American Chemical Society)

Depending on the generation number, the number of end-groups per dendron is increasing from one amine group in the linear polymer to 8 amine groups in the third generation dendron. The end groups were additionally modified with maltose. With increasing generation number, not only do the number of end-functionalities increase; but volume of the single dendrons increases too. Taking into account the fact that almost 90% of the monomer units in the linear backbone are dendronised, it can be expected that with increasing generation this structure should have huge change on the conformation of the single DenPols. Hence, MD simulations showed that the linear backbone (MI-G0-Mal) exists in a helical conformation with rather short and thick chains, as additionally confirmed by atomic force microscopy. Modification with dendrons in the first and second generations leads to disturbance of the helical conformation and the formation of coil-like conformations. In contrast, modification with the third generation results in thin, approximately 500 nm long worm-like chains. This is a result of the very dense packing of the third generation dendrons along the linear chain which leads to stretched macromolecules with a very dense distribution of the maltose functionalities along the linear backbone. AF4 measurements of the molar masses of the resulting species indicate very well controlled aggregation behavior depending on the generation number, and on the conformation of the DenPols, respectively. The well-defined, rather stiff structures of the G0 and G3 DenPols carry maltose functionalities exposed for further interaction and self-assembly, leading to pronounced pH-dependent aggregation. In contrast, the coil-like conformation of G1 and G2 shields significant amount of the maltose units, which are not available for further interactions and show limited aggregation (Figure 5a).

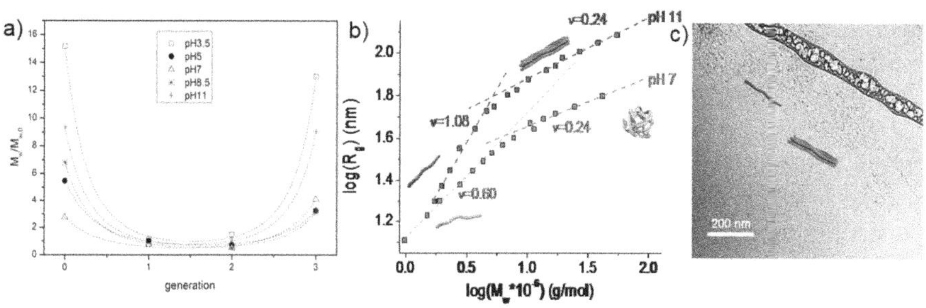

Figure 5. a) Aggregation number $M_w/M_{w,0}$ of the different generations DenPols MI-G0-Mal−MI-G3-Mal, depending on the pH,b) Dependence of the R_g on the molar mass of MI-G3-Mal at pH11 and 7, and c) Cryo TEM images of MI-G3-Mal at pH 11, large structures in image belongs to TEM grid. (reprinted with permission from Boye et al. Biomacromolecules 2012, 13 (12), 4222-4235, copyright (2012) American Chemical Society)

AF4 coupled to a static light scattering detector obviously enables a very clear statement about the aggregation under different pH conditions by calculating the molar mass of the aggregates. However, much more detailed information could be achieved by the combination of the radius of gyration and the molar mass both determined using light scattering detection after the AF4 separation. Based on the scaling law (equation 1), the conformation plot gives the scaling parameter from the different components in system. In case of DenPols, single macromolecules and their self-assemblies were indicated, showing a distinct difference in the scaling exponents calculated for the single macromolecules and for the self-assemblies. For example, the third generation DenPol exists at pH 11 as rigid rods ($\nu \sim 1$), which interact to form laterally composed self-assemblies ($\nu \sim 0.2$) (Figure 5b). This behavior and the worm-like structure was impressively confirmed by cryoTEM measurements as shown in Figure 5c. When the pH is changed to 7, immediately the single chain becomes more like a flexible coil while the self-assemblies possess a sphere-like structure. It is noteworthy that the complementary combination of AF4-LS with microscopic investigation for this study is of immense importance, since similar scaling values are valid for different special assemblies, which can be realized only with the help of additional TEM studies. However, vice versa, microscopy alone can only deliver discrete impression and not a full picture of all components existing in this quite complex system and their size and conformation distribution.

Interestingly, pH has a pronounced influence on all generations of the single DenPols and their self-assemblies and hence a 3D dependency of their conformation on concentration, pH and generation number was established based on orthogonal AF4-LS measurements.

Understanding of the interaction mechanism, conformation and size of the different generation DenPols was additionally a valuable information in understanding their behavior when exposed to other particles. The worm-like structure in combination with huge amount of maltose units are available for further interactions, especially in the third generation DenPol. These key characteristics deliver the basis for supracolloidal interactions with cellulose nanocrystals (*53*). Cellulose nanocrystals possess similar length (~500 nm) as the stretched DenPols and a number of functionalities for H-bonding interactions with the maltose units of the DenPol (Figure 6). This leads to supracolloidal structures in which the DenPol wraps along the cellulose nanocrystal, as observed by transmission electron microscopy and tomography. Such interactions are of significant interest when developing novel, sustainable materials with tunable properties (*54–56*).

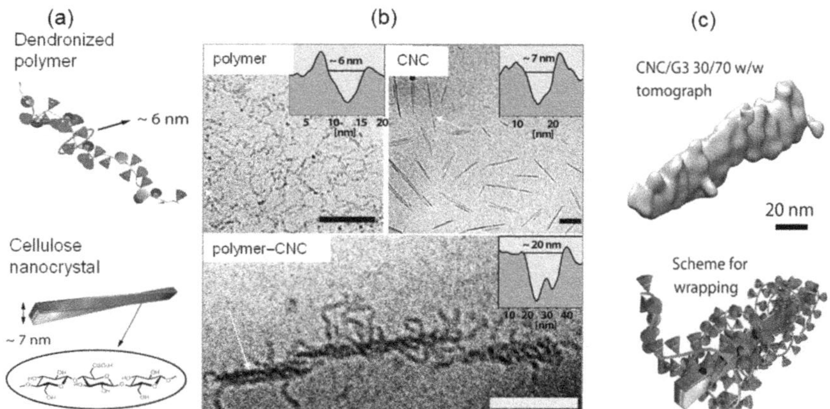

Figure 6. a) Schematic illustration of G3 and Cellulose nanocrystal (CNC), b) Cryo HR-TEM images of the DenPoLs, the CNC and the polymer-CNC complex, scale bars are 100 nm, and c) ET image polymer-CNC complex and the related scheme. (reprinted with permission from Majoinen et al. J. Am. Chem. Soc. 2014, 136 (3), 866-869, copyright (2015) American Chemical Society)

In applications as biosensors, in the biomedicine, or in the nanoelectronics, single conjugates define the active processes. Therefore, understanding of these biomacromolecular structures at the molecular level is essential for effective design of such materials. AF4 can deliver much needed information on the size, conformation and quantity of bioconjugates and their distributions. Avidin-biotin interaction is often used to deliver biofunctional supramolecular structures (*57,*

58). This interaction is non-covalent but very strong and selective since only four biotin units can interact with the four free valances of the avidin. Hence, controlled modification of poly(propylene imine) glycodendrimers (*21*) with biotin units can lead to selective interaction with avidin. This results in discrete bioconjugate structures (Figure 7a).

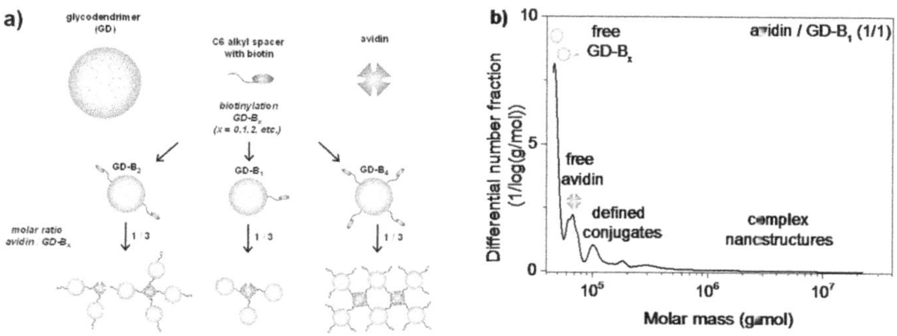

Figure 7. a) Schematic overview of initial components (glycodendrimer (GD), spacered biotin ligand (B), and avidin) and idealized examples for the formation of biohybrid structures by differently biotinylated glycodendrimers, and b) molar mass distribution of bioconjugates formed by avidin and monovalent biotynilated GD (GD-B$_1$) in equimolar ratio (1/1) with schematic structure assignment. (reprinted with permission from Boye et al. Macromolecules 2015, 48 (13), 4607-4619, copyright (2015) American Chemical Society)

Although DLS and microscopy studies can give first impression about the development of bioconjugates depending on the chosen ratios (*19*, *20*), both methods are limited in terms of size and conformation distributions. DLS in batch delivers only an average value over the molecular sizes and does not contain any structural information. The limits of microscopy are based on the fact that only descrete fractions of the whole sample can be quantitatively analyzed. Complete quantitative and qualitative information in a broad size range can be delivered by AF4 coupled to static and dynamic light scattering. The size of the single components is in the range of 40 – 60 kg/mol while differently structural bioconjugates can possess molar masses of around 104 kg/mol. An AF4 separation shows the single components, bioconjugates and large networks in a single separation run (*22*). Quantification of their amount in the system is simply performed based on concentration detection with differential refractometer (Figure 7b). The qualitative interpretation of the formation and the conformation of the conjugates is possible using the light scattering and calculation of the molar masses,the radius of gyration and hydrodynamic radius. The three parameters deliver powerful analysis of the complete conjugate diversity, which obviously depend on the number of biotin units per dendrimer as well as the ratio between avidin and biotinylated dendrimer in the system. These investigations show simultaneously various key issues: low molar mass, well defined bioconjugates in large amount, formation of short, stiff and linear supramolecular structures, branched structures at ~ 10^6 g/mol and the formation of networks with ~10^7 g/mol. The extraction of these information from the AF4 measurements is possible by the combination of a conformation plot (Equation 1) together with the calculation of the ρ-parameter (Equation 3, Figure 8) and the apparent density (Equation 2).

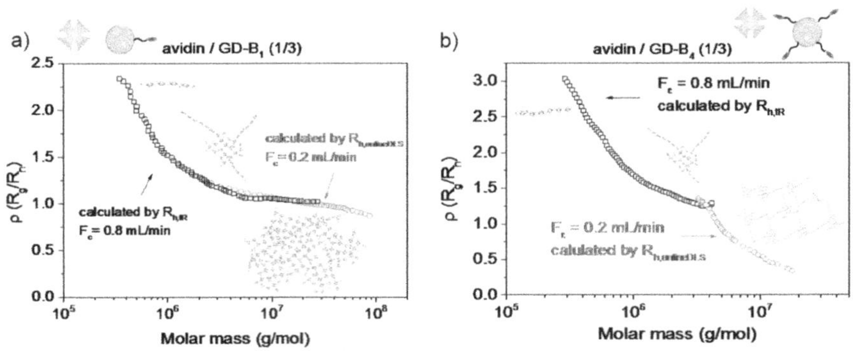

Figure 8. ρparameters vs molar mass obtained by AF4 coupled to static and dynamic light scattering detection of biohybrid structures formed by a) avidin/GD-B_1 (1/3) and b) avidin/GD-B_4 (1/3). (reprinted with permission from Boye et al. Macromolecules 2015, 48 (13), 4607-4619, copyright (2015) American Chemical Society)

It is worth mentioning, that the detection of the hydrodynamic radius of these large distributions by using DLS is onlyapplicable for sizes lower than 50 nm. For larger conjugates the fact that AF4 separation is based on different diffusion properties of particles withdifferent sizes, the retention properties in AF4, which after quantitative interpretation, can be used for calculation of the hydrodynamic radius (*28–31*). The calculation of the hydrodynamic radius from the retention properties of the particles through the AF4 channel turns to be a more suitable approach for larger particles. This is not surprising, since the two approaches are based on different detection of the diffusion constant. For DLS the time dependent fluctuation of the measured light intensity is needed, and for large particles with slow diffusion, longer measuring times are needed. However measurements in flow offer limited time window for the DLS approach. This problem does not exist in the case of the calculation of the diffusion coefficient from the retention time.

Finally, the interpretation of the ratio between the hydrodynamic and the radius of gyration together with the apparent density gives essential information about the network formation of biotinylated glycodendrimer and avidin for their further application. While dendrimers carrying two biotin units form dense, globular network, increase of the biotin units per dendrimer leads to the formation of low density, loose networks.

The suitability of AF4 for the characterization of large biomacromolecular systems is straight forward based on the separation approach and the application of light scattering detectors. However, when evaluating biomacromolecular systems, the question of their interaction with small drug molecules as a guest in a large macromolecular system is of significant interest (*59*). Both, AF4 and static light scattering are limited in the separation and detection of small molecules. In AF4, the membrane and the frit, needed for performing the cross-flow function of the separation channel, and the cut-off of the membrane are the limiting factors, allowing for molecules smaller than the cut-off to pass through it. Thus, small molecules cannot be eluted through the channel to reach the detectors, even if they would be suitable for detection. This filtration effect of the AF4 membrane is often used for purification of samples (*60, 61*). Yet, it could be useful also for the separation of small particles and their quantitative analysis in macromolecular hybrid system (Figure 9).

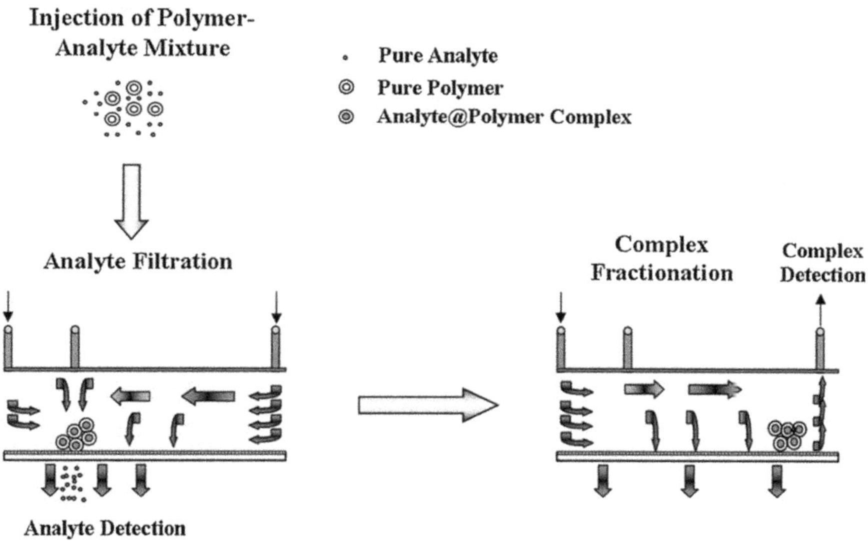

Figure 9. Principle of AF4 separation method for analyte filtration and analyte@polymer complex fractionation. (Reprinted from Journal of Chromatography A, 48, Boye et al., An alternative route to dye-polymer complexation study using asymmetrical flow field-flow fractionation, 4607-4619, Copyright 2010, with permission from Elsevier)

In this particular example, a hyperbranched polymer with a poly(ethylene imine) scaffold and a maltose shell was investigated as a host for small drug molecules (*24*). This polymer has a weight average molar mass of approx. 200 000 g/mol. As a model for a drug, the dye Rose Bengal was used, as it is not expensive as a test substance and, additionally, can be easily detected using UV-Vis absorption. Rose Bengal has a molar mass of approx. 1000 g/mol, which is smaller than the membrane cut-off of 5000 g/mol. The effect of encapsulation of Rose Bengal from the hyperbranched glycopolymer was observed at low pH, while release takes place at high pH (Figure 10).

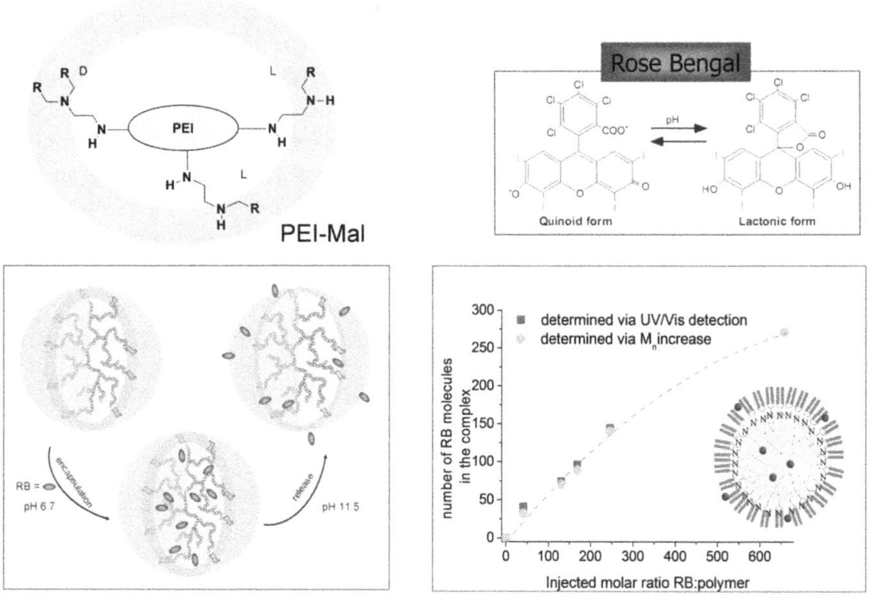

Figure 10. Complexation studies of the host molecule PEI-Mal (a) and the guest molecule Rose Bengal (b), schematic principle of pH sensitive encapsulation and relase (c) and results of quantitative determination of complexed Rose Bengal amount by AF4 studies (d). (Reprinted from Journal of Chromatography A, 48, Boye et al., An alternative route to dye-polymer complexation study using asymmetrical flow field-flow fractionation, 4607-4619, Copyright 2010, with permission from Elsevier)

However, quantification of the encapsulated amount in the polymeric macromolecules was not easy to perform by UV-Vis measurements because of the chemical complexity of the macromolecular system. Therefore, AF4 was applied for quantitative separation and detection of the encapsulated Rose Bengal amount in two simultaneous ways:

(i) Collecting the non-encapsulated Rose Bengal during the focusing step after the injection of the mixture of hyperbranched polymer and dye, in which the polymer-dye complex is already formed. A UV-Vis detector is coupled to the cross-flow outlet, which delivers the quantity of the separated free dye after calibration with defined dye concentrations.
(ii) Standard AF4 separation and light scattering detection procedure of the polymer-dye complex remaining in the channel after the focusing step delivers the molar mass of the complex.

Different ratios between hyperbranched glycopolymer and dye enable setting up a dependence between the molar ratio in the batch and the calculated amount in the complex after separation of the free dye. Additionally, the molar mass of the complex increases with increasing amount of dye in the batch. These two dependences overlap very well showing encapsulation up to 144 Rose Bengal molecules per hyperbranched macromolecule.

The potential of this method to quantify encapsulation efficiency is obviously high. However, the limitations of this technique should be considered when developing a quantitative analysis of complex systems. For example, the interaction between the single components with the membrane can strongly affect the results. Usually, regenerated cellulose is used in the AF4 channel, which should be carefully tested before quantification of the results. In the case of Rose Bengal such interaction was observed, though membrane modification with additional Rose Bengal led to saturation of free valences for interaction and delivered a stable layer allowing for reliable interpretation of the data. Furthermore, focusing time and injection amount could lead to aggregation or destruction of the conjugates and should be subject of careful evaluation in order to avoid misinterpretations.

The investigation of biomacromolecules by AF4 with different detection techniques is a powerful tool to obtain information of sizes and conformations in large size distributions. The quantification of the single components and, thus, the evaluation of conjugation and encapsulation processes are highly reliable. In fact, this separation approach and the interpretation of the detector data requires in-depth understanding of the fractionation mechanisms and the effort required for developing suitable method for every particular biomacromolecular system is comparably high. However, the information achieved after a single injection is of superior quality, taking into account, that with very small amounts of bioactive substances, essential amount of details about their molecular properties and their distributions are available. Thus, AF4 has huge potential in the analysis of complex biomacromolecular architectures for applications in which interaction with natural materials are of interest.

References

1. Pasch, H.; Trathnigg, B. *Multidimensional HPLC of Polymers*; Springer-Verlag: Berlin Heidelberg: 2013.
2. Radke, W. Polymer separations by liquid interaction chromatography: Principles – prospects – limitations. *J. Chromatogr. A* **2014**, *1335*, 62–79.
3. Meunier, D. M.; Smith, P. B.; Baker, S. A. Separation of Polymers by Molecular Topology Using Monolithic Columns. *Macromolecules* **2005**, *38* (12), 5313–5320.
4. Gerber, J.; Radke, W. Topological separation of linear and star-shaped polystyrenes by off-line 2D chromatography. Stars having high molar mass arms and quantification of the star fraction. *Polymer* **2005**, *46*, 9224–9229.
5. Edam, R.; Eeltink, S.; Vanhoutte, D. J. D.; Kok, W. T.; Schoenmakers, P. J. Hydrodynamic chromatography of macromolecules using polymer monolithic columns. *J. Chromatogr. A* **2011**, *1218*, 8638–8645.
6. Al Samman, M.; Radke, W.; Khalyavina, A.; Lederer, A. Retention Behavior of Linear, Branched, and Hyperbranched Polyesters in Interaction Liquid Chromatography. *Macromolecules* **2010**, *43*, 3215–3220.

7. Ahn, S.; Lee, H.; Lee, S.; Chang, T. Characterization of Branched Polymers by Comprehensive Two-Dimensional Liquid Chromatography with Triple Detection. *Macromolecules* **2012**, *45*, 3550–3556.
8. Wang, F.; Lu, C.-H.; Willner, I. From Cascaded Catalytic Nucleic Acids to Enzyme–DNA Nanostructures: Controlling Reactivity, Sensing, Logic Operations, and Assembly of Complex Structures. *Chem. Rev.* **2014**, *114*, 2881–2941.
9. Villalonga, M. L.; Díez, P.; Sánchez, A.; Gamella, M.; Pingarrón, J. M.; Villalonga, R. Neoglycoenzymes. *Chem. Rev.* **2014**, *114*, 4868–4917.
10. Tu, Y.; Peng, F.; Adawy, A.; Men, Y.; Abdelmohsen, L. K. E. A.; Wilson, D. A. Mimicking the Cell: Bio-Inspired Functions of Supramolecular Assemblies. *Chem. Rev.* **2016**, *116*, 2023–2078.
11. Sinkeldam, R. W.; Greco, N. J.; Tor, Y. Fluorescent Analogs of Biomolecular Building Blocks: Design, Properties, and Applications. *Chem. Rev.* **2010**, *110*, 2579–2619.
12. Sapsford, K. E.; Algar, W. R.; Berti, L.; Gemmill, K. B.; Casey, B. J.; Oh, E.; Stewart, M. H.; Medintz, I. L. Functionalizing Nanoparticles with Biological Molecules: Developing Chemistries that Facilitate Nanotechnology. *Chem. Rev.* **2013**, *113*, 1904–2074.
13. Leonard, A.; Dandoy, P.; Danloy, E.; Leroux, G.; Meunier, C. F.; Rooke, J. C.; Su, B.-L. Whole-cell based hybrid materials for green energy production, environmental remediation and smart cell-therapy. *Chem. Soc. Rev.* **2011**, *40*, 860–885.
14. Guix, M.; Mayorga-Martinez, C. C.; Merkoçi, A. Nano/Micromotors in (Bio)chemical Science Applications. *Chem. Rev.* **2014**, *114*, 6285–6322.
15. Gaitzsch, J.; Huang, X.; Voit, B. Engineering Functional Polymer Capsules toward Smart Nanoreactors. *Chem. Rev.* **2016**, *116* (3), 1053–1093.
16. Bünzli, J.-C. G. Lanthanide Luminescence for Biomedical Analyses and Imaging. *Chem. Rev.* **2010**, *110*, 2729–2755.
17. Apostolovic, B.; Danial, M.; Klok, H.-A. Coiled coils: attractive protein folding motifs for the fabrication of self-assembled, responsive and bioactive materials. *Chem. Soc. Rev.* **2010**, *39*, 3541–3575.
18. Boye, S.; Appelhans, D.; Boyko, V.; Zschoche, S.; Komber, H.; Friedel, P.; Formanek, P.; Janke, A.; Voit, B. I.; Lederer, A. pH-Triggered Aggregate Shape of Different Generations Lysine-Dendronized Maleimide Copolymers with Maltose Shell. *Biomacromolecules* **2012**, *13*, 4222–4235.
19. Ennen, F.; Fenner, P.; Stoychev, G.; Boye, S.; Lederer, A.; Voit, B.; Appelhans, D. Coil-like Enzymatic Biohybrid Structures Fabricated by Rational Design: Controlling Size and Enzyme Activity over Sequential Nanoparticle Bioconjugation and Filtration Steps. *ACS Appl Mater Interfaces* **2016**, *8*, 6261–8.
20. Ennen, F.; Fenner, P.; Boye, S.; Lederer, A.; Komber, H.; Voit, B.; Appelhans, D. Sphere-Like Protein-Glycopolymer Nanostructures Tailored by Polyassociation. *Biomacromolecules* **2016**, *17*, 32–45.
21. Ennen, F.; Boye, S.; Lederer, A.; Cernescu, M.; Komber, H.; Brutschy, B.; Voit, B.; Appelhans, D. Biohybrid structures consisting of biotinylated glycodendrimers and proteins: influence of the biotin ligand's number and

chemical nature on the biotin–avidin conjugation. *Polym. Chem.* **2014**, *5*, 1323–1339.

22. Boye, S.; Ennen, F.; Scharfenberg, L.; Appelhans, D.; Nilsson, L.; Lederer, A. From 1D Rods to 3D Networks: A Biohybrid Topological Diversity Investigated by Asymmetrical Flow Field-Flow Fractionation. *Macromolecules* **2015**, *48*, 4607–4619.
23. Yassin, M. A.; Appelhans, D.; Wiedemuth, R.; Formanek, P.; Boye, S.; Lederer, A.; Temme, A.; Voit, B. Overcoming Concealment Effects of Targeting Moieties in the PEG Corona: Controlled Permeable Polymersomes Decorated with Folate-Antennae for Selective Targeting of Tumor Cells. *Small* **2015**, *11*, 1580–1591.
24. Boye, S.; Polikarpov, N.; Appelhans, D.; Lederer, A. An alternative route to dye-polymer complexation study using asymmetrical flow field-flow fractionation. *J. Chromatogr. A* **2010**, *1217*, 4841–9.
25. Wahlund, K. G.; Giddings, J. C. Properties of an asymmetrical flow field-flow fractionation channel having one permeable wall. *Anal. Chem.* **1987**, *59*, 1332–9.
26. Ratanathanawongs Williams, S. K.; Lee, D. Field-flow fractionation of proteins, polysaccharides, synthetic polymers, and supramolecular assemblies. *J. Sep. Sci.* **2006**, *29*, 1720–1732.
27. Moquin, A.; Winnik, F. M. The Use of Field-Flow Fractionation for the Analysis of Drug and Gene Delivery Systems. In *Field-Flow Fractionation in Biopolymer Analysis*; Williams, S. K. R., Caldwell, K. D., Eds.; Springer Vienna: Vienna, 2012; pp 187−206.
28. Moon, M. H.; Giddings, J. C. Size distribution of liposomes by flow field-flow fractionation. *J. Pharm. Biomed. Anal.* **1993**, *11*, 911–920.
29. Nilsson, L.; Leeman, M.; Wahlund, K.-G.; Bergenståhl, B Mechanical Degradation and Changes in Conformation of Hydrophobically Modified Starch. *Biomacromolecules* **2006**, *7*, 2671–2679.
30. Magnusson, E.; Håkansson, A.; Janiak, J.; Bergenståhl, B.; Nilsson, L. Hydrodynamic radius determination with asymmetrical flow field-flow fractionation using decaying cross-flows. Part II. Experimental evaluation. *J. Chromatogr. A* **2012**, *1253*, 127–133.
31. Håkansson, A.; Magnusson, E.; Bergenståhl, B.; Nilsson, L. Hydrodynamic radius determination with asymmetrical flow field-flow fractionation using decaying cross-flows. Part I. A theoretical approach. *J. Chromatogr. A* **2012**, *1253*, 120–126.
32. Burchard, W. Solution Properties of Branched Macromolecules. In *Branched Polymers II*; Roovers, J., Ed.; Springer Berlin Heidelberg: Berlin, Heidelberg, 1999; pp 113−194.
33. Nzé, R.-P.; Nicolai, T.; Chassenieux, C.; Nicol, E.; Boye, S.; Lederer, A. Effect of Connectivity on the Structure and the Liquid–Solid Transition of Dense Suspensions of Soft Colloids. *Macromolecules* **2015**, *48* 7995–8002.
34. Mai, T.; Rakhmatullina, E.; Bleek, K.; Boye, S.; Yuan, J.; Völkel, A.; Gräwert, M.; Cheaib, Z.; Eick, S.; Günter, C.; Lederer, A ; Lussi, A.; Taubert, A. Poly(ethylene oxide)-b-poly(3-sulfopropyl methacrylate) Block

Copolymers for Calcium Phosphate Mineralization and Biofilm Inhibition. *Biomacromolecules* **2014**, *15*, 3901–3914.
35. Mai, T.; Boye, S.; Yuan, J.; Volkel, A.; Grawert, M.; Gunter, C.; Lederer, A.; Taubert, A. Poly(ethylene oxide)-based block copolymers with very high molecular weights for biomimetic calcium phosphate mineralization. *RSC Adv.* **2015**, *5*, 103494–103505.
36. Haladjova, E.; Rangelov, S.; Geisler, M.; Boye, S.; Lederer, A.; Mountrichas, G.; Pispas, S. Asymmetric Flow Field-Flow Fractionation Investigation of Magnetopolyplexes. *Macromol. Chem. Phys.* **2015**, *216*, 1862–1867.
37. Erber, M.; Boye, S.; Hartmann, T.; Voit, B. I.; Lederer, A. A convenient room temperature polycondensation toward hyperbranched AB2-type all-aromatic polyesters with phenol terminal groups. *J. Polym. Sci., Part A: Polym. Chem.* **2009**, *47*, 5158–5168.
38. Voit, B.; Appelhans, D. Glycopolymers of Various Architectures—More than Mimicking Nature. *Macromol. Chem. Phys.* **2010**, *211*, 727–735.
39. Taori, V. P.; Lu, H.; Reineke, T. M. Structure–Activity Examination of Poly(glycoamidoguanidine)s: Glycopolycations Containing Guanidine Units for Nucleic Acid Delivery. *Biomacromolecules* **2011**, *12*, 2055–2063.
40. Spain, S. G.; Gibson, M. I.; Cameron, N. R. Recent advances in the synthesis of well-defined glycopolymers. *J. Polym. Sci., Part A: Polym. Chem.* **2007**, *45*, 2059–2072.
41. Richards, S.-J.; Jones, M. W.; Hunaban, M.; Haddleton, D. M.; Gibson, M. I. Probing Bacterial-Toxin Inhibition with Synthetic Glycopolymers Prepared by Tandem Post-Polymerization Modification: Role of Linker Length and Carbohydrate Density. *Angew. Chem., Int. Ed.* **2012**, *51*, 7812–7816.
42. Rawat, M.; Gama, C. I.; Matson, J. B.; Hsieh-Wilson, L. C. Neuroactive Chondroitin Sulfate Glycomimetics. *J. Am. Chem. Soc.* **2008**, *130*, 2959–2961.
43. Okada, M. Molecular design and syntheses of glycopolymers. *Prog. Polym. Sci.* **2001**, *26*, 67–104.
44. Imberty, A.; Chabre, Y. M.; Roy, R. Glycomimetics and Glycodendrimers as High Affinity Microbial Anti-adhesins. *Chem. Eur. J.* **2008**, *14*, 7490–7499.
45. Geng, J.; Biedermann, F.; Zayed, J. M.; Tian, F.; Scherman, O. A. Supramolecular Glycopolymers in Water: A Reversible Route Toward Multivalent Carbohydrate–Lectin Conjugates Using Cucurbit[8]uril. *Macromolecules* **2011**, *44*, 4276–4281.
46. Dai, X.-H.; Dong, C.-M.; Yan, D. Supramolecular and Biomimetic Polypseudorotaxane/Glycopolymer Biohybrids: Synthesis, Glucose-Surfaced Nanoparticles, and Recognition with Lectin. *J. Phys. Chem. B* **2008**, *112*, 3644–3652.
47. Chabre, Y. M.; Roy, R., Chapter 6 - Design and Creativity in Synthesis of Multivalent Neoglycoconjugates. In *Advances in Carbohydrate Chemistry and Biochemistry*; Horton, D., Ed.; Academic Press, 2010; Vol. 63, pp 165–393.
48. Atilano, M. L.; Pereira, P. M.; Yates, J.; Reed, P.; Veiga, H.; Pinho, M. G.; Filipe, S. R. Teichoic acids are temporal and spatial regulators of

peptidoglycan cross-linking in Staphylococcus aureus. *PNAS* **2010**, *107*, 18991–18996.

49. Ahmed, M.; Narain, R. The effect of polymer architecture, composition, and molecular weight on the properties of glycopolymer-based non-viral gene delivery systems. *Biomaterials* **2011**, *32*, 5279–5290.
50. Lederer, A.; Burchard, W. *Hyperbranched Polymers: Macromolecules in-between of Deterministic Linear Chains and Dendrimer Structures*; Royal Society of Chemistry, 2015.
51. Svenson, S.; Tomalia, D. A. Dendrimers in biomedical applications—reflections on the field. *Adv. Drug Delivery Rev.* **2012**, *64*, 102–115.
52. Gillies, E. R.; Fréchet, J. M. J. Dendrimers and dendritic polymers in drug delivery. *Drug Discovery Today* **2005**, *10*, 35–43.
53. Majoinen, J.; Haataja, J. S.; Appelhans, D.; Lederer, A.; Olszewska, A.; Seitsonen, J.; Aseyev, V.; Kontturi, E.; Rosilo, H.; Osterberg, M.; Houbenov, N.; Ikkala, O. Supracolloidal multivalent interactions and wrapping of dendronized glycopolymers on native cellulose nanocrystals. *J. Am. Chem. Soc.* **2014**, *136*, 866–9.
54. Lin, N.; Huang, J.; Dufresne, A. Preparation, properties and applications of polysaccharide nanocrystals in advanced functional nanomaterials: a review. *Nanoscale* **2012**, *4*, 3274–3294.
55. Klemm, D.; Kramer, F.; Moritz, S.; Lindström, T.; Ankerfors, M.; Gray, D.; Dorris, A. Nanocelluloses: A New Family of Nature-Based Materials. *Angew. Chem., Int. Ed.* **2011**, *50*, 5438–5466.
56. Habibi, Y.; Lucia, L. A.; Rojas, O. J. Cellulose Nanocrystals: Chemistry, Self-Assembly, and Applications. *Chem. Rev.* **2010**, *110*, 3479–3500.
57. Thordarson, P.; Le Droumaguet, B.; Velonia, K. Well-defined protein–polymer conjugates—synthesis and potential applications. *Appl. Microbiol. and Biotechnol.* **2006**, *73*, 243.
58. Cosnier, S. Biomolecule immobilization on electrode surfaces by entrapment or attachment to electrochemically polymerized films. A review. *Biosens. Bioelectron.* **1999**, *14*, 443–456.
59. Liechty, W. B.; Kryscio, D. R.; Slaughter, B. V.; Peppas, N. A. Polymers for drug delivery systems. *Annu. Rev. Chem. Biomol. Eng.* **2010**, *1*, 149–73.
60. Li, P.; Giddings, J. C. Isolation and measurement of colloids in human plasma by membrane-selective flow field-flow fractionation: Lipoproteins and pharmaceutical colloids. *J. Pharm. Sci.* **1996**, *85*, 895–898.
61. Giddings, J. C.; Yang, F. J.; Myers, M. N. Flow Field-Flow Fractionation Channel as a Versatile Pressure Dialysis and Ultrafiltration Cell. *Sep. Sci.* **1977**, *12*, 499–510.

Editors' Biographies

Yongmei Wang

Yongmei Wang (Ph.D) is a professor in Chemistry Department at The University of Memphis. She received her B.S. degree from University of Science and Technology of China in 1985 and PhD in Chemistry from University of Notre Dame in 1990. Her research interests are theoretical/computational studies for polymer characterization, polymer based gene delivery and nanomedicines.

She has published over 80 peer-reviewed journal articles, organized/co-organized symposiums in the areas of separation and characterization of macromolecules and nanoparticles, gene delivery and nanomedicines. She was the recipient of Dunavant Professorship from the University of Memphis, and the Alumni Association of Distinguished Research Award from the University of Memphis.

Wei Gao, Ph.D.

Wei Gao (Ph.D) is a Senior Research Scientist in Analytical Sciences, Corporate R&D, at The Dow Chemical Company. She received her B.S. degree from Fudan University in 1991 and PhD in Polymer Chemistry and Physics from Peking University in 1997. Her research interests are on synthesis-structure-property relationship of polymeric and colloidal products, characterization of macromolecules and colloids through separation with advanced detections, and sustainable polymers.

She has published over 30 peer-reviewed journal articles and book chapters, more than 20 patents and patent applications. She organized /co-organized ACS symposiums in the areas of separation and characterization of macromolecules and particles, polymer sciences for everyday things, and polymer colloids. She also served in scientific program committee for GPC2015 conference and the international symposium on Field- and Flow-based separations (FFF2018). She was the recipient of NOVA Innovation Award from Rohm and Haas Company, and US EPA Presidential Green Chemistry Challenge Awards.

Sara Orski, Ph.D.

Sara Orski is a Research Chemist in the Materials Science and Engineering Division at the National Institute of Standards and Technology (NIST). She earned her B.S. degree at The College of William and Mary in 2006 and her Ph.D. in Chemistry at The University of Georgia in 2011. Her current research interests are in the design, synthesis, and characterization of regularly branched polyolefins, which can be used as model materials to quantify the effect of macromolecular chemistry and architecture on dilute solution properties, and can serve as potential next-generation standard reference materials for the US plastics industry.

She has published over 20 peer-reviewed journal articles, 4 book chapters, and a patent. She is an active in scientific community service including serving as an ACS councilor and the chair of the Younger Chemists Committee (YCC) for her local ACS section, The Chemical Society of Washington. She is also co-chair of the ACS POLY-PMSE Student Chapters organizing committee.

X. Michael Liu

X. Michael Liu, Ph.D. is a Global Associate Research Fellow at Pfizer, Inc. He serves as an editorial board member for *International Journal of Analytical Mass Spectrometry, Chromatography and Chromatographic Science*, and *Application of Chromatography in Pharmaceutical Analysis*. Dr. Liu is an accomplished scientific leader with strong functional management experience. He received a Ph.D. in Chemistry from University of Massachusetts, a M.S. in Product Development from Rochester Institute of Technology, and a B.S. in Biochemistry from Sun Yat-sen University. He has extensive knowledge in product development, pre-formulation, material characterization, and analytical chemistry. With over 20 years of industrial experience, Dr. Liu has actively involved in developing pharmaceutical and medical device products for global markets, building and leading technical teams, securing regulatory approvals, and supporting commercial efforts.

Dr. Liu has published more than 25 peer-reviewed journal articles, 20 patents and patent applications. He organized/co-organized technical symposia focused on polymers in biomedical/pharmaceutical applications and characterization in national ACS meetings. His current research interests include preformulation of coatings and soft-gels used for various pharmaceutical dosage forms. He also helped create innovative technology platforms to support new product development and process improvements.

Indexes

Author Index

Subject Index

A

B

D

G

I

www.ingramcontent.com/pod-product-compliance
Ingram Content Group UK Ltd.
Pitfield, Milton Keynes, MK11 3LW, UK
UKHW020816290926
15470UKWH00003B/1200

9 780841 233096